CISO Impact and Influence

CISO Impact
AND Influence

How Cybersecurity Executives
Can Take the Lead and
Nudge the World

Chris Brown

New Cyber Press
PORTLAND, OR

Published by New Cyber Press
Portland, Oregon
media@newcyberexecutive.com
newcyberexecutive.com

New Cyber Press books are available at special quantity discounts for bulk purchase for sales promotions, events, fundraising, and educational needs. Special books or book excerpts also can be created to fit specific needs. For details and permission requests, write to the email address above.

ISBN 979-8-9928826-1-2 (eBook)
ISBN 979-8-9928826-0-5 (paperback)
ISBN 979-8-9928826-2-9 (hardback)

VI.0

—

Book Midwifery by Fen Druadin
Copyediting by James Gallagher
Proofreading by Adeline Hull
Illustrations by Maria Malecha
Author Photo by Luke Copping
Book Design & Publishing by Kory Kirby
SET IN GARAMOND PREMIER PRO

TO THE TEACHERS AND ADULTS OF MY CHILDHOOD—*those who shared wisdom, sparked curiosity, and left me with insights that linger far beyond the classroom.*

TO THE EXTRAORDINARY SET OF FRIENDS FROM GRADE SCHOOL—*the kind of bond so rare and unlikely that most wouldn't believe it existed. But it did, and it mattered.*

TO THE MANY MANAGERS, LEADERS, AND MENTORS IN MY ADULTHOOD—*those who guided, challenged, and shaped my growth in ways both intentional and unspoken.*

TO THE COLLEAGUES AND CLIENTS WHO LISTENED WITH INTEREST AND PATIENCE AS I DEVELOPED AND REFINED IDEAS—*your engagement gave my thoughts form and meaning.*

TO ALL WHO READ THIS BOOK—*may you be the shoulders upon which those who follow you stand, just as those who came before me did for me.*

This is for all of you.

Contents

Introduction　　　　　　　　　　　　　　　　　　　*1*

Part I: Mindsets

Chapter 1: Impactful Cybersecurity Leadership Mindsets
　　Are Business Mindsets　　　　　　　　　　　　　11
Chapter 2: Know When to Rely on Familiar Framings and
　　When to Create New Ones　　　　　　　　　　　25
Chapter 3: Mindsets Set You Apart　　　　　　　　　45

Part II: Relationships

Chapter 4: Ground Yourself in the Wider World;
　　Skip the Window Dressing　　　　　　　　　　　61
Chapter 5: Business Influence Develops from Genuine Interest in
　　Others' Concerns and Goals　　　　　　　　　　77
Chapter 6: Strategy Creates Its Own Self-Calibration
　　and Momentum　　　　　　　　　　　　　　117

Part III: Development

Chapter 7: Purpose Drives Subtle Yet Powerful Moves That Guide You
　　and That Others Can Embrace　　　　　　　　147
Chapter 8: Practicing with Today's Small Issues Prepares You for
　　Tomorrow's Big Challenges　　　　　　　　　169
Chapter 9: Sustainable Growth Thrives on Steady Mechanisms　183

Conclusion　　　　　　　　　　　　　　　　　　*191*

Appendix

A: Questions to Move from Concepts to Action   195
B: The Real Budget Process   201
C: Involving Others in Security Responsibilities   207
D: Contributing to Group Discussions on Topics
   You Are Unfamiliar With   213
E: Learning About Your Peers Through Associations   217
F: Learning About Your Peers Through Relationships   221
G: Developing Your Organization's Board of Directors   223
H: Implicit Bias Dimensions Among CISOs   231
I: Building Consensus   233

Index   *239*
About the Author   *245*

Inscription

"Today I escaped from the crush of circumstances,
or better put, I threw them out, for the crush wasn't
from outside me but in my own assumptions."
—MARCUS AURELIUS, *Meditations*, 9.13

When we are low, or when the world feels particularly hard, we often point to external circumstances as reasons for not achieving our goals.

But the truth—the freedom we often avoid—is that no matter how low we feel, we can shape our world. We always have choices.

The way we frame our circumstances is a choice.

The interpretations we give our feelings are a choice.

Our perspectives are choices.

Our mindsets are choices.

These choices are the essential work of leadership in our field. As an executive, making choices is your responsibility—and the path to the impact only your choices can create.

Introduction

"Cybersecurity is in trouble."

The way most people think about cybersecurity is wrong. And unless we correct our thinking, we put our companies, our communities, and the very foundations of our democracy at risk.

Cybersecurity is no longer an isolated concern.

As cybersecurity executives we have the potential to influence the world well beyond what we typically think of as our domain.

This book emerges from a deep desire to help maximize that potential. I'm worried because our key leaders—the cybersecurity executives responsible for shaping our approach—are being disrupted in their success.

Critical risks remain unaddressed by both industry and government. Chief information security officers (CISOs) stand at a crossroads, bearing the brunt of risks that could upend not only their own organizations but society as a whole.

We as leaders shoulder work that can be hard, frustrating, and isolating. We often feel like we're fighting for those who don't see or appreciate what we do. But, as cybersecurity leaders, we have the power to change this—to reshape our paths and redefine the field itself.

Invitations come our way all the time, but we don't always see them or take advantage of them. This book is an invitation to you. You, as a cybersecurity leader, have an opportunity to create the circumstances for a better world—for yourself, for those who rely on you, and for those who will follow in your footsteps.

So if you're along for the ride, buckle up. Together we'll face the brutal realities of executive leadership in cybersecurity, because the right approach transforms isolation and risk into influence, fulfillment, and ease. This isn't a book to stroke egos or indulge in platitudes about best practices; it's a wake-up call for CISOs—a prompt to meet the role's strategic demands head-on.

"You have to know how to put it back together."

The summer before kindergarten, my parents got me a sandbox and Tonka toy excavator. I loved that excavator and played with it every chance I had. After a while I realized *I'm going to get bored of this soon.* I could feel the fun fading. So I decided to figure out how it worked.

I sneaked a screwdriver from the garage and took the excavator apart, but I couldn't put it back together, and no kid I asked had ideas that worked for my situation. So instead, I carefully stacked the pieces so they appeared to still be assembled.

Eventually my dad found out and said, "Chris, if you want something from the garage, come talk to me first. And if you're going to take something apart, know how to put it back together. Playing with things in ways you don't fully understand can get you hurt—or cause real damage."

I learned the limits of what I could do alone, as well as the limits of what even a group of smart five-year-olds could do and the advantages of getting help beyond other five-year-olds (or CISOs).

Later, as I transitioned from technical expert to business executive, I learned how to benefit from mentorship and collaboration, building relationships and drawing from the wisdom and insight of others outside my field.

"I want to nudge the world."

I didn't do homework as a kid. I mean, it was legendary. Across the school, before I walked into new classrooms, teachers and fellow students already knew Chris Brown didn't do homework—even in the advanced classes I somehow ended up in year after year. This reputation stuck with me from kindergarten through high school.

That is, until . . .

. . . my senior year AP physics teacher pulled me into an administration office. He said, "We're here because you're not doing homework, but let's not talk about that. What do you want to do with your life?"

I thought about it in silence for a long time.

He patiently waited.

Finally, I said, "The whole universe interacts. It's a set of interconnected events, all moving in different and ever-changing ways. My hope is to nudge it just a little bit, in a slightly better direction."

He replied, "You're in a better place to do that if you do your homework."

He made sense of otherwise senseless homework, and he tapped into my intrinsic motivation—probably the first time I'd been coached instead of told, encouraged, scolded, punished, or bribed. Someone had finally connected something I didn't want to do with something I wanted to achieve.

That's one purpose of this book—to tap into your intrinsic motivation to tackle things you don't want to but will anyway because you connect those things to a bigger motivation. CISOs understand intrinsic motivation well, and it's a powerful force for us to tap into.

While not every part of this book will work for every person or every situation, everything here can be adopted in some form. My hope is that you see possible actions, behaviors, and concepts that help you put yourself in a better position to nudge the world—or whatever your goal might be.

I share "soft" ideas with real-world impact. The ideas can create bigger outcomes for those willing to step back, see the big picture, and adjust. These ideas and tools have been tested by me, by the bosses who shared them with me, and by my consulting and executive coaching clients. Their simplicity is their strength, and they drive change that's both important and hard to measure.

If that's not for you, please pass this book along to someone who may benefit from it.

My Writing Intent

Many books prescribe the author's way of doing things. In contrast, I intend to help you discover new ways to think about your circumstances, tackle challenges, and improve over time. My approach encourages you to let go of weak assumptions and to change behaviors and habits rather than acquiring knowledge. It's about transforming how you think about *how* you think, not simply adding to what you know.

Plan to Make a Big Impact; Stay Long Enough to See It Through

Research shows that 30 percent of executives in new roles don't last six months. Half don't make it past two years. Fifty percent of CISOs are considering leaving the field altogether.

Those numbers are bad for individual cybersecurity executives, bad for their staff and organizations, bad for those who follow us, and bad for the field as a whole.

Imagine instead a world where cybersecurity executives find fulfillment, support, and recognition as standout leaders.

This vision excites me—helping cybersecurity executives in stressful, ill-defined roles (often with low authority and big responsibilities) to move into zones of higher success, lower stress, and harmony between work and personal life.

In short, the pages of this book offer what I do when coaching CISOs and cybersecurity executives.

If this book empowers even a single reader to find a better path, leading to personal realignment or personal transformation, I'll consider it a success. If it reaches two or more, then perhaps I've helped shape culture across organizations, and these leaders might create ripples that transform the field further.

Maybe I'll have nudged the world.

How to Use This Book

Star athletes spend more time visualizing their plays than they do on the court or field. This book invites you to reflect similarly—on the position you hold now, or the position you aspire to—and to envision your future challenges along with potential setbacks and successes. As you are exposed to the ideas here, consider your past actions and behaviors in light of these new ideas. Think about what would have worked for you, and how you might have adapted it for your style, preferences, and strengths.

The approaches here are meant to be both immediately useful and an inspiration for new ideas relevant to your needs. They offer a structure for navigating human challenges and fostering continuous learning and adaptation. Once you've internalized these principles, then orienting, reframing, and calibrating will come naturally, no matter the environment.

This book introduces metacognitive methods, new mindsets, and tools rarely discussed among cybersecurity leadership. These pages offer ideas, but your engagement will turn these ideas into effective tools.

So slow down whenever you feel resonance or discomfort; both are signs that something deeper is at work. Slowing down at the right moment is a gift. Don't refuse that gift.

Throughout, I'll use this clock to prompt you to pause at valuable times:

At the end of each chapter, you'll also find questions that bring awareness to your thinking patterns, habits, and defaults. These questions help you uncover what is holding you back—and what can move you forward.

> **Note:** If you want to return to your thoughts, jot your answers to questions and prompts in the margins or in your favorite notebook.

What's Next

In part 1, we'll examine assumptions and mindsets that hold you back from the careers, recognition, influence, and satisfaction you desire. In part 2, I will help you build meaningful relationships with peers and leaders, and connect meaningfully around shared goals. In part 3, we'll focus on continued development of thinking processes, comprehensive growth, and self-coaching as a platform for long-term success.

Questions

1. Where are you still operating hands-on? Is it habit, comfort, or preference—and is it holding you back from fully stepping into your executive role?
2. Looking back, when did seeing the bigger picture make a difference?
3. What impact do you want to make in the world, and what does that look like?

PART I:

Mindsets

CHAPTER 1:

Impactful Cybersecurity Leadership Mindsets Are Business Mindsets

"Think of the children!"

You've probably heard a variation of this in a movie or in a debate.

Does it work? Does it sound a bit melodramatic and overdone?

How about this?

"Because cyber!"

For most non-cybersecurity executives, "because cyber"—in its many forms—sounds just as melodramatic and overdone, not to mention hollow and repetitive. Yet time and again the rationale for cybersecurity—in pitches for budget, resources, or attention—boils down to "Think of the children!" Or rather: "Because cyber!"

This mindset claims that the value of cyber is self-apparent.

But it isn't self-apparent.

And this becomes our first example of mindset in action—one that holds us back.

Nothing is "because cyber."

This is circular reasoning and a faux shortcut. Without significant groundwork, it's also lazy.

Even when the chief executive officer (CEO) supports cyber with this sentiment, we'd be wise to not let them get away with it for long.

At some point, its power wears out.

Let's do better.

"How does that help my business?"

One of my first jobs during college was as a programmer and system administrator for a local business. I convinced the owner to invest in a permanent internet connection—a T1 line. This was a costly operational expense for a twenty-five-person company, but he liked something about my pitch, and we got the T1.

A few months later I suggested a firewall might be a good idea.

He looked at me—clearly frustrated—and said, "Why do I care about a firewall? I sell cash registers. How does that help my business?"

I said, "Because security!"

No, I didn't, actually.

I stopped. I considered the question.

I let it change how I thought about security.

That one question shifted my career. I began asking myself why, trying to turn tacit knowledge of security into articulable risk to the business. Now, as then, business leaders don't want technical benefits; they want to see how security contributes to their success.

Over the decades, my everyday focusing question has been refined to "What can cyber do for the business?"

1.1 Stay on the Right Side of Credibility to Maintain Authority

Cyber leaders focus to a great extent on technology and reducing perceived technical risk. Far too little attention is given to business acumen and pragmatic relevance.

In my conversations with non-cyber leaders about their CISOs, no one—truly no one in leadership—says, "I wish my CISO was more technical."

I repeatedly hear the following refrains: "I want them to understand the business," "I want them to work with their peers," and "I want them to focus on the business with a cyber lens."

It's table stakes for today's cybersecurity executives to answer their local version of the questions "Why do I care about a firewall? How does it help me sell cash registers?"

1.2 Bring Your Unique Strengths, but Treat Cybersecurity as "Just Business"

Cybersecurity is nothing more than what a company is already worried about, just manifesting in a way that CISOs are charged with paying attention to. Put another way: *Cybersecurity is simply one view of risks that a business already faces.*

Despite the spotlight on cybersecurity at the board level, its frequent separation as a category in enterprise risk, and its singular treatment by cybersecurity executives, there's no first-principle (or most fundamental) reason to separate cybersecurity from other risks.

Solutions are often first-order (immediate), and while a second-order (long-term) reason for isolating cyber risk is to manage its volatility, uncertainty, complexity, and newness, applying clear first-principle thinking is always more powerful.

Working from first principles, treating cybersecurity as a "special" risk brings its own problems and undermines our goals. The more we treat cybersecurity as a distinct entity, the more we'll find ourselves isolated as leaders and cut off from the business we're supposed to be integral to—even while carrying responsibilities we can't possibly address on our own.

Mental models shape our behavior, thinking, and decision-making. So how effective does the *separate and special* approach strike you?

Embracing the idea that cybersecurity is simply part of business—while allowing for distinct conversations—helps tackle one of cybersecurity's biggest challenges: aligning business goals with cyber's purpose and meaningful action.

Here's a default for many organizations: tagging a known-exploited third-party software vulnerability on an internal website with "cybersecurity"—effectively turning an operational maintenance shortcoming into something for cybersecurity to solve. That creates bad (perverse) incentives everywhere, and, frankly, most of us know this. If not explicitly, then instinctively.

In bypassing the first-principles, we create seemingly intractable problems. The "cybersecurity" risk bucket lumps together distinct problems with different origins, unique solutions, and impacts on different value chains.

Most of what we do as CISOs is to try to undo this first-principle mistake.

What if we changed that?

If we *don't* treat cyber as its own thing, we change our strategic field. We shift from threat actors, technology constraints, and budget caps to holistic thinking about business concerns of non-cybersecurity executives: optimizing profits, preventing losses, and balancing opportunity costs and risk.

1.3 Calibrate Cybersecurity Resources by Linking Them to Business Opportunities

Businesses care about risk only in relation to opportunity—expressed in opportunities measured in market share, shareholder value, operational excellence, and delivery on mission. This opportunity mindset of business executives drives business success.

Generally speaking—aside from regulations—a cybersecurity label is irrelevant from a business lens. What matters is where each risk fits into the organization's opportunities and the risk-for-opportunity bets involved.

Cybersecurity resources will be apportioned based on their perceived protection of things business executives already care about.

We see evidence of this when CISOs rely on broad statements like "This risk could impact the brand" to convey importance. But these clichés often fail to connect with real impact. Blanket statements—based on assumptions like "bad news equals brand damage"—often reflect CISOs operating outside their lane of expertise.

If the chief marketing officer couldn't (or wouldn't) finish your sentence with "and that would result in brand damage," then the CISO shouldn't even attempt to finish the sentence.

Even if you aren't saying these clichés out loud, *are you using them as a placeholder in your thoughts?* If so, how does that affect your credibility as an executive?

Here's a challenge: How does your cybersecurity mental model change if you adopt the following statement?

> Cybersecurity risks are a filtered view of business risks, not a distinct bucket of risk. There are few, if any, pure cybersecurity risks. There are few, if any, risks that are for the CISO to assess or decide on alone.

This should make you slightly uncomfortable. But discomfort is where growth occurs. Lean into it.

If it doesn't make you uncomfortable, lean into it until it does.

Say to yourself, "What if there are no risks for me to address as a unitary leader? None. Not explicitly or implicitly. Risk decisions always involve another leader outside of cyber."

It's extremely rare that a CISO takes this far enough.

I'll give you a minute—because if you take this far enough, it changes everything about how you show up as a business leader.

When you stop treating cybersecurity risks as isolated issues and start seeing them as intertwined with business opportunities, everything changes. You shift from being a lone risk mitigator to a business leader helping the organization to make better, bolder decisions.

1.4 Creative, Acceptable, and Practical Solutions Emerge from the Business

Early in my executive career, an IT manager working on the disaster recovery (DR) plan assumed that restoring payroll would be important during a regional disaster, as employees rely on pay continuity for stability during a crisis. This was interesting and insightful. In the draft plan, payroll was scheduled for recovery within the first three days of a disaster.

During a review of the plan with the four most-senior executives—as a matter of luck—the human resources (HR) leader was present.

Upon hearing the plan, she said, "We don't need that module that quickly. We have our checks cut by a major payroll company that's not likely to be directly operationally affected by a regional disaster. We can instruct them to cut the same payroll as last cycle, and fix what we need to with a phone call. What we really need is a phone."

Poof! With a single decision, the recovery of Accounts Payable—an enterprise resource planning (ERP) module notorious for being difficult to recover early in DR—was instantly deprioritized from a top priority to a position in the low thirties of the recovery order. As a result, overall DR complexity decreased by about 10 percent.

This could be figured out only in a business context. The same is true for all technical risks. A proper understanding of options and priorities can happen only with a deep understanding of the business.

Note: "Technical risk in a business context" and "operational risk in a business context" follow a recursive pattern. For example, an attack that randomly alters a few unused Domain Name System (DNS) entries is a risk relevant to IT. An attack targeting DNS entries for externally hosted HR systems, however, becomes a risk to the HR function's business activities and mandates. Similarly, an attack affecting DNS entries for corporate banking websites is now a risk for the finance function. That specific financial risk, in turn, may pose a broader risk to the entire business. Generally, risks remain the responsibility of the level directly impacted. However, in specific cases—such as when broader business operations or outcomes are jeopardized—they transform into concerns for higher organizational levels. In short, business context is defined as what is determined as relevant by those outside the risk carrier's immediate domain.

The only people who have that understanding are in the business. As cybersecurity professionals we can't count on ourselves to know enough to contextualize risk or make decisions on behalf of people whose jobs and responsibilities we have only a surface understanding of.

The problem isn't understanding how to solve the technical problem; the problem is understanding the business context for the technical problem.

1.5 Unease Motivates, but Protecting Opportunity Compels

Risk exists only in the context of opportunity. A list of risks alone is unanchored from business goals and easy to dismiss. But when risks are framed in a business-specific context, each risk gains meaning in light of the opportunities and sources of value to be protected.

Leaders use the following framework: *Is this opportunity worth the risk?* It encompasses the logic, instinct, and emotion in every choice, balancing possibilities against potential consequences.

Consider these risks:

- skin scrape or minor cut (medium likelihood)
- paper cut (medium likelihood)
- quick death (low likelihood)
- lost limb (very low likelihood)
- stolen wallet (low likelihood)

Now think of these situations:

- finally climbing Half Dome
- filing paperwork at the office
- going on a dream vacation with your family
- working a heavy manufacturing job to support your family
- receiving treatment for an aggressive, debilitating, painful, and eventually life-ending disease

Context is everything—you don't need numbers or math to know which risks are worth it—and depending on how you pair the risks and the opportunities, a reasonable response could range from "Hell, yes!" to "Hell, no!"

The same applies to cybersecurity risk conversations. Without grounding in real business opportunities, discussions devolve into the use of heat maps that substitute artificial constructs—like color gradient—to compensate for a lack of real business context.

> **Note:** Notice that I didn't need to provide "impact" estimates. Describing the scenarios conveys them well enough. Trust your peers and leaders to make sensible opportunity–risk decisions, then frame it and communicate only the pieces you have that they don't.

1.6 Models That Ignore Business Fundamentals Are Fundamentally Flawed

Consider this example: A risk quant estimates a $20 million cyber-risk tolerance for a $1 billion company. That tolerance reflects the company's threshold for loss without affecting core objectives. The CISO keeps cyber risk under this threshold for a year, with little pushback—or understanding—from business leadership.

Then business leadership starts increasing its cyber-risk tolerance. The CISO feels blindsided, imagining disaster scenarios and pondering whether to get personal legal advice. It later comes to light that this increase in tolerance was intentional in the lead-up to a market play expected to boost revenue by 60 percent and profits by 180 percent, with low incremental cybersecurity exposure. Therefore, even a 100 percent increase in risk exposure in the near term might be business sensible, if not compelling. The risk was more than covered by opportunity.

This is the reality of business: Risk and opportunity go together.

The greater the distance we put between an opportunity and its associated risks, the harder it is to justify addressing those risks—for the business and cybersecurity alike.

Thinking about risks without understanding opportunity kills influence and authority—every single time.

Leverage Technical Models, but Don't Showcase Them

Most of the models a CISO uses to communicate with the business are right for the CISO's direct reports to manage cybersecurity but actually wrong for the CISO as a business executive, because these models tend to increase the distance between business opportunity and cyber risk.

There's a reason control and risk frameworks fail, technical threat models lack relevance, and risk management for cybersecurity's sake doesn't get traction.

There's also a reason you almost never hear about Generally Accepted Accounting Principles (GAAP) in finance, even though GAAP determines much of what finance does in accounting and how it does it. The same goes for the models that specialists across the business rely on but rarely mention.

If a model doesn't have direct business relevance for your stakeholders, don't show them.

Bluntly, it's not enough to be technically right; you must be situationally right.

Career-wise, if your models are wrong, you may hear this about your performance:

- You lack executive presence.
- You need to think strategically.
- You're not ready for an executive role.
- You're not a team player.

Professionally, if your models are wrong, you may:

- be disappointed by the reactions of your boss and internal business partners.
- receive a lack of input in settings where others get it.
- be frustrated by low-quality professional feedback.
- experience confusion about performance.
- have uncertainty about job security.
- (in the extreme) be significantly affected by frustration, resentment, and burnout.

So if technical models are a poor basis for connecting cybersecurity to the business, what will work?

Talking to people.

Just do that.

Seriously. With the right information and relevant framing, people often know what to do.

1.7 Don't Let Appearances Limit Your Approach or Influence

Most defined and episodic business processes have informal and ongoing processes that occur in parallel. This is something that all executives are expected to leverage.

Let's illustrate with a specific example.

The default budgeting process for most cybersecurity programs goes like this: (1) the cybersecurity department puts in a request for budget, (2) cyber receives a set budget, let's say, pegged to something like a percentage of IT, and (3) cyber uses what they get to address what they view as the most critical technical risks.

But that's not how budgeting works. Annual budgeting is a multiyear process—even if it doesn't appear so, and even when budget managers and finance insist it happens only during specific periods as part of a distinct cycle.

Rather than accept the limited view, consider what you can do to advocate for budget:

1. Understand the official process and its pitfalls and hidden complexities.

2. Understand the parallel/political/informal process and its nuances.

3. Build relationships with decision influencers and decision-makers.

4. Understand the value of cyber initiatives and operations to decision influencers.

5. Iterate and refine the budget requests and value propositions/justification with their input.

6. Informally shop your plan to decision influencers and decision-makers.

7. Adjust your plans based on new insights on interests, concerns, and imperatives.

8. Submit budget request paperwork.

Only #8 is truly once a year. The rest are ongoing.

Remember the story about homework putting me in a better position to nudge the world? That's what we're talking about here. Steps 1–7 are your homework, and they put you in a better position to nudge the world (protect or increase your budget).

Recognize that your organization will have several key processes or practices—not just budgeting—that appear one-dimensional or episodic, and it's part of your job as an executive to figure out whether the appearance is the full story and to capitalize on the power and influence of relationships, which have outsize influence on how "logic" and "necessity" are determined.

(For more on budgets, see appendix B.)

Questions

1. Where do you feel misaligned or disconnected from the larger organization?
2. As you reflect on the role of the CISO, what past situations—whether personal experiences or observations of other leaders—might you reconsider in light of the leadership, strategy, and communication challenges discussed in chapter 1?
3. When do your cybersecurity explanations come across as "Because cyber!" to non-cybersecurity executives? What will you need to improve this?

CHAPTER 2:

Know When to Rely on Familiar Framings and When to Create New Ones

"I don't care what you say, we're not doing it."

At one stage in my career, I moved from being an IT operations and security manager into a role as a senior consultant with a distinguished boutique health-care management consulting firm—my first consulting role. One of my initial engagements involved delivering Health Insurance Portability and Accountability Act (HIPAA) security training to a large regional hospital's nursing staff. It was my first experience with user training in a paid engagement.

My colleague led the first session while I observed. This guy was a heavy hitter in the field, one of the first *real* HIPAA Security Rule consultants. Smart on information security and the regulation—and not the kind of consultant who spelled it HIPPA.

I delivered the next session on my own. It fell flat. Afterward, my colleague explained that I hadn't explained what the purpose was—why anybody in the room should care. He also said, "You also slipped and said *customer*. Don't do that again. It's *patient*."

An hour later I delivered the next session.

At the same point as the earlier session, the nurses were again energetically explaining why they had to remove the screen protectors—this was before nurses' stations were reconfigured for privacy—and the whole idea of privacy and security was relatively new.

This time a woman in the middle of the classroom said, "The screen protectors slow us down and get in the way of providing patient care. I don't care what you say. I'm going to continue to rip the screen savers off, and so will everybody else."

She then went on to explain that she was the chief nursing officer (CNO).

<Expletive>.

Despite a firm commitment from the CEO and chief information officer (CIO) to comply with regulations, the CNO was literally saying, "No, not going to do this."

I had no doubt that she would win that battle, or, if she didn't win outright, that her opponents in the fight would walk away with a metaphorical scar or two—and possibly a permanent limp. She had that kind of authority.

I didn't have a snowball's chance in hell of arguing with her directly.

I moved on and continued the training. But I also started to listen, because my colleague had primed me and the CNO had given the gift of scaring me enough to set aside *my agenda* to *listen to her*, as well as to the comments and concerns from the nurses in the room.

Listening, it hit me that the problem had a possible answer: patient care.

In one of those flashes that I am still thankful for to this day, I had a glimmer of an idea, and even before developing where it would lead, I asked:

"When do you stop caring about patient care?"

The CNO, who had been politely listening to me deliver the training, now shifted.

"What do you mean?" she asked.

"Do you stop caring about patients after discharge?"

"No, we want them to go home, recover, get well, and stay well."

"What if a patient had a situation they were ashamed of, and a neighbor visiting someone else was walking the halls, saw this patient's information, and went home and blabbed it to the neighbors? You discharge the patient, and when they get home, the neighbors are talking about the patient, asking questions. The patient gets stressed, stops participating in the community, and ends up avoiding follow-up care out of shame and embarrassment."

A long pause.

"The screen protectors will stay on the monitors," said the CNO with the same authority from earlier.

I knew every nurse in the hospital was going to hear about what happened in that room and was going to be on board with at least the screen protectors—and likely a whole lot more.

This situation challenged me and changed me.

My central concerns about cybersecurity became centered on one question: How do I reach this audience?

This was similar to the question my former boss asked, "Why do I care about a firewall?"

But this was deeper.

It was more fundamentally human.

It was emotional, tied to personal values, and an entire innate and cultivated set of passions and beliefs.

It tied to a worldview and a self-identity.

If you've never worked with a CNO, they are some of the most hard-ass people on the planet. I'd put even odds on a battle of wills between a CNO and an elite Special Forces marine.

Both do their jobs because they care so much about what they do, and they are going to do what they think is right. Get them on your side and you have an ally like no other.

Seeing the CNO change her position over the course of that fifty-minute training session created something in me—a spark of possibility—and an understanding that *the way to reach people is to connect to something they already deeply care about.*

"Everyone is an owner, but not really."

In my first role as CISO, I was tasked with what seemed, on the surface, to be a straightforward problem: managing identity and access more efficiently. Like many companies, ours had an access review process that required each manager to verify access rights for everyone on their team.

Seems logical, right?

It was needlessly unorganized, as there were no baselines, and every manager was checking every aspect of access against each person, including the overwhelmingly complex claims adjudication system.

But the real problem—the first-principle problem—was that supervisors shouldn't be deciding access. Those calls belong to the process champions or data stewards—the people who understand the critical larger organizational impacts, information flows, and stakeholder expectations. But at that time I didn't have the language for it. I couldn't articulate it in a way that would resonate across the board.

So I sat down with the corporate strategy director, hoping to get some clarity. She listened, then did something unexpected. She took out a thesaurus, placed it on her desk between us, and opened it. Together we explored language that would clarify roles and responsibilities around access, data, and process authorities. We needed language that was precise and in plain English. We were looking for the words that would make these roles self-evident, that would render the purpose of access review and how it fit into larger concepts comprehensible across the organization.

When we landed on those terms, everything changed. Access management transformed from an "IT and infosec thing" to a framing that anyone could understand. The language cut through the noise and created a common understanding. And here's the kicker: None of it was wrapped up in cybersecurity jargon. That moment taught me the power of language—not just any language, but language that resonates beyond the technical.

We didn't just fix a process; we created a foundation for understanding and clarity. We carved out a structure around access that, even now, has stayed with me as one of the simplest but most powerful lessons in aligning cybersecurity with business in a way that simply works.

Note: Owner: Reserved solely for the company or the data subject. To prevent ambiguity and confusion, ownership is not assigned to any individual leader within the company.

Application Sponsor: The individual with the greatest vested interest in a specific application, or who would experience the most significant business impact if the application were not in place.

Data Steward: The designated person who oversees decision-making related to data processing, usage, and sharing for specific types or categories of information. Typically this is associated with the department responsible for data intake of that specific data.

Process Champion: The individual accountable for the entire business process (not a single procedure) within their domain, such as the member-enrollment process, encompassing all associated subprocesses and activities.

Custodian: The IT, business analysis, or other operational support teams responsible for maintaining, managing, and supporting the underlying data, infrastructure, tools, and technical functions of the process.

2.1 Technical Knowledge Alone Won't Build Authority with Peers and Leaders

Our professional authority with others is shaped by multiple factors. It begins with expertise in our own field but expands as we orient ourselves toward the disciplines and perspectives of stakeholders.

Language plays a central role in this and serves as a starting point for connection and as a path to deeper understanding. When we approach business and specialist language with genuine curiosity and respect, it becomes a tool for learning about other fields, broadening our perspectives, and building honest connections. Daniel Kahneman's "fast thinking" and "slow thinking" concepts from *Thinking, Fast and Slow* apply here: Quick rapport is the early benefit, but over time people feel appreciated when others show genuine interest in subjects they care about. As your understanding grows, stakeholders will increasingly view you positively.

Just as finance, sales, and marketing have their own "languages," companies have organizational terms tied to culture, self-conception, and history. Familiarity with these nuances signals that you "get" the organization, and this creates an immediate sense of shared purpose. For instance, one of my clients uses "Northstar" to embody a sense of unified vision, values, and direction, all in a single word. It is a powerful term that can do a lot of heavy lifting when used with understanding and discernment.

In cybersecurity, we can default to using our specialized language, implicitly or explicitly expecting others to adapt. Instead, aim to reverse this approach. Go beyond "plain English" and "simple concepts" and understand and adopt the language of others—understand their professional values and ethics. This is, after all, the same respect you'd hope they'd show you and your field.

2.2 Don't Demand Authority; Build Influence from Where You Are

In my first role as a CISO, I had the opportunity to report to three exceptional leaders over the span of six years: the CIO, the chief financial officer (CFO), and the chief risk officer (CRO). This gave me firsthand insight into the distinct functional approaches within the same company and culture, all while keeping my similar responsibilities as a CISO. I learned which factors were shaped by individual leadership styles, by distinct functional perspectives, and by company culture. A key takeaway from this experience was that in a high-caliber organization—indeed, in most organizations—the title of your supervisor matters less than the strength of your relationship with them and how you lead from your current position, wherever in the organization that might be.

While I leaned into the opportunities within each of those reporting relationships, I've seen many CISOs spend their time waiting for more authority, a bigger title, or a different line of reporting, because they believe it will help them achieve their goals. However, they often become frustrated when those expectations aren't met—sometimes to the point of leaving.

That's a shame, because it's based on one of the myths of executive leadership: that there's a level where you suddenly gain magical "authority."

The reality is far different. The higher up you go, the more interdependent you are on other executives. In most environments, influence significantly replaces authority. Influence is how you have impact at the top, even for CEOs.

Exceptional leaders take responsibility for what they can and use the resources and relationships available to them.

Lead from where you are.

> **Note:** If the CEO doesn't support your preferred reporting structure, it's for one of three reasons: You haven't made a compelling case, you haven't demonstrated readiness, or it doesn't make business sense. Determining which of these might be in play is tricky, to put it mildly, but figuring this out (chapters 4 and 5) gives you a path forward.

2.3 Readiness and Progression Is About Your Next Role, Not This One

When working with clients on promotions, I use a Lilliputian-on-giant-stairs metaphor to explore the concept of next-level readiness. Picture yourself as a Lilliputian—a tiny figure—standing in front of giant stairs that represent career progression. This is especially helpful for interim CISOs vying for the CISO role, but it applies to everyone.

The mindset that underlies this visualization is that readiness is not being good at what you do; readiness is being good at the next thing.

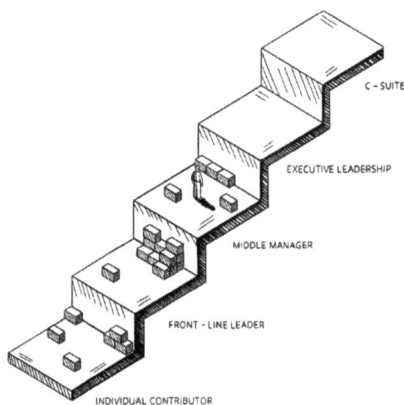

The Readiness Staircase™ is a five-step staircase, each step representing a level of management, with steps significantly taller than the person climbing them. From lowest to highest, the steps are labeled: "Individual Contributor," "Front-Line Leader," "Middle Manager," "Executive Leadership," and "C-Suite." Progression through management levels requires more than simply standing on a step—it depends on readiness, illustrated by the height of the stacks of blocks above current role expectations. Taller stacks indicate greater preparedness to advance, while shorter stacks suggest less.

Standing directly on the stairstep start means you're executing at your current level—potentially very well—but haven't taken on any next-level work yet. Striving for the next level while starting from the flat—a danger zone—represents non-readiness for the next level.

Attempting promotion from a flat start carries high risks, putting strain on both you and those who rely on you, including your sponsor and the organization. Despite what you may think, you don't want to be promoted from these positions. The risk of significantly underperforming in the new role is too high.

Promotion from a moderate height stack of next level work (the challenge zone) is achievable with support, while promotion from a tall stack of next level work (the easy zone) is often less about readiness and more about the right opportunity. The ideal promotion occurs when you're consistently operating somewhere the challenge zone and the easy zone.

If you've started taking on informal next-level work, you're on a small stack—like a Lilliputian jumping to reach the next stair—stretching hard to grab the next step, sometimes missing and needing to try again. With just one next-level project, the effect of a miss isn't catastrophic.

As you take on more next-level work, you move into the challenge zone, where the gap to the next step is smaller and the effort less intense.

By the time you're on stack approaching the next level, significant next-level work is already part of your portfolio. Moving to the next step still takes effort, but it's rarely a struggle, and you rarely miss.

The goal is to increase elevation consistently—taking on next-level work until you're confidently operating on the green slope, ready for a full-time move to that next step.

This is promotion readiness—the result of building skills and proving you're prepared for what comes next.

2.4 Likability Opens Doors, but Going as a Team Gets Approval

You will likely be entertained as a CISO when offering proposals or ideas to an executive team or a decision-maker such as the CEO. After all, you progressed from cyber professional into the most-senior cyber executive role for a reason.

That said, at the level of a CISO, expectations change. As with any role making a pitch, preparation is expected. But as an executive, that preparation now includes involving key stakeholders, gathering insights, adjusting, and garnering support or avoiding strong resistance.

If you don't do this homework, don't expect the CEO or another decision-maker to support you. These executives have to work together, and no one *wants* to be in a position supporting you if it means clashing with a powerful leader or burning political capital.

It's part of your job to provide grief-avoidance—minimizing friction, frustration, and unnecessary obstacles for other leaders—whenever you can, and especially when you're seeking their support. While you may not be able to eliminate these entirely, showing that you've tried goes a long way.

Involving other stakeholders is also a powerful way to build trust in you as an executive, all around.

2.5 Shape Plans as a Team and Go with the Team

The following are common but often unrecognized pitfalls in developing support for an initiative or an ongoing operation in a corporate environment:

- Pitching with limited perspective.
- Pitching without allies.
- Pitching without political guidance.
- Pitching without addressing uncertainty.

Early cybersecurity leaders often overlooked these by default, to a level that lingers in people's minds today. Even experienced cybersecurity executives can inadvertently fall back into these pitfalls.

The "limited perspective" and "without allies" pitfalls can manifest in many ways but are fundamentally a knowledge asymmetry. One manifestation occurs when a dissenting voice in a group, such as on a committee, can halt an entire process because they knew something you didn't, or had a significant set of business questions that you were not prepared for. Compounding this, if you haven't done your homework, the potential allies in the room—even frequent allies—don't know enough to back you up, or even whether they should.

The "without political guidance" pitfall is when your idea or proposal gets in the way of someone's agenda and that person quietly and persistently derails you—or, in rare cases, does so loudly and paints you as incompetent or green.

The "without addressing uncertainty" pitfall occurs when the CISO feels confidence where others don't. It manifests as a response from decision-makers, such as "We need more information," "Did you talk to so-and-so?" or "We'll have to decide this after _____" (often an unrelated event from which there's a reduction in overall organizational uncertainty and "room" for the uncertainty you're trying to introduce).

In most cases, these are on you for not doing groundwork, and they are all preventable.

A simplified example:

"Let's do next-gen firewalls; it's only going to be $1.5 million, and it will protect us better," says the CISO.

While cybersecurity professionals might understand the value provided by next-gen firewalls and have a sense for the timeline of implementation and time to value, your audience might not. Now put that up against the following:

"Let's do a $2 million marketing initiative that will increase sales by 10 percent while reducing sales cost by 5 percent in the next nine to twelve months."

It's easy to see who will get more enthusiastic approval in a room that's hearing both proposals for the first time. *CISOs also have more homework to do than most executives because they are opportunity preservers versus creators.*

Bring Other Executives into the Process to Overcome These Pitfalls

Consider the tactic of a child saying "Mom said I could" to get Dad to go along.

Invoking "Mom said I could" works for children for two reasons. First, it prompts the audience, Dad, to consider any real objections rather than quickly dismissing the idea based on natural skepticisms, instinct, habit, source, or lack of initial information. It gives you time. Second, it gives your idea credibility due to the trusted source. It lets you borrow credibility from someone else, even when they are not present.

Now consider a later-in-childhood, more-advanced technique: "Bobby's mom said she would pick me up and stay while Bobby and I play games, and then drop me off later. Mom said she was okay with that, but to double-check with you to make sure you didn't need me for anything."

This example has several additional facets: (1) the demonstrated planning and responsibility encompassed in anticipating objections, (2) the clearing of objections before they have a chance to solidify, (3) showing that the initiative has external support and is likely to achieve specific goals ("safety" in the case of the child), and (4) framing the type of objections Dad could focus on. It still borrows credibility, but this time from both Mom and another parent.

This same mechanism applies in business.

To give this a more grown-up handle to allow us to refer to the approach, I'll call it "going as a team" for convenience.

Do Your Homework with Your Peers to Avoid Poor Assumptions and Fundamental Errors; Then Show Your Work to Go as a Team

How do you get approvals and support for a cybersecurity project?

Just check with your mom, and then say, "My mom said I could."

Not really, of course. But the idea of borrowing credibility can be adapted by presenting information in a way that effectively invokes teams or trusted people, even in their absence.

This signals being a team player, understanding implications from a broad company perspective, and, of course, having a team that supports the success of the undertaking—again, whether or not they are physically present. The process has extended your credibility and the attractiveness of your idea.

In a business context, an illustrative example might be stated as "Finance reviewed this and said it met their threshold criteria for getting funding." This is the "Mom said I could" version.

A more advanced version might be "I discussed this with finance. They suggested a few changes to part 6 on how I presented the ROI, but said it was a strong proposal, that it should be funded, and it would be approved without issues in purchasing. Several product teams liked it and asked when we could start." Such detailed communication about interactions with departments preemptively addresses the most common types of objections.

This also reduces cognitive load for the listener. They no longer have to operate on behalf of the finance department, nor guess how much political capital it would take to make the approval happen in finance—assuming the decision-maker wants to support it—because you already did that work. Multiply that over several departments, and the decider (CEO, etc.) can now focus on the decision strategically, as a CEO rather than administratively, or from multiple perspectives.

The other significant advantage of this process is that you're priming the discussion and decision, if it comes up when you are not present. You have essentially created a level of organizational advocacy among all the people you have spoken with.

2.6 Trust Isn't One Thing—It's Many

Trust is a requirement in business—but trust itself is complex. We trust others for many reasons: because they keep their word, show consistency, protect confidentiality, collaborate well, and support others' growth, to name a few.

While most understand you need trust to get things done, not everyone notices that there are many things we trust others for (or don't). Knowing which applies when, recognizing when each is needed, and, importantly, not lumping them into one category of "trust" is a better starting point for building trust with others.

• To Keep Their Word	• To Communicate Openly and
• To Be Fair	Effectively
• To Tell the Truth	• To Show Respect for Boundaries
• To Respect Confidentiality	• To Demonstrate Competence
• To Act with Integrity	• To Be Responsible
• To Be Dependable in Crisis	• To Support Others' Growth
• To Make Ethical Decisions	• To Share and Collaborate
• To Show Consistency	• To Show Empathy and
• To Be Loyal	Understanding
• To Work Collaboratively	• To Act with Humility

Examples of types of trust.

Understanding the type of trust required in a given relationship—whether with peers, direct reports, or executives—provides a clear advantage in building and maintaining that connection. Similarly, knowing the forms of trust your peers value enables you to strengthen these relationships by demonstrating the trust they need from you, both in your role and your interactions.

Think about the kinds of trust that are important to you.

What might be important to others?

Your family and friends?

Your direct reports?

Your leaders and peers?

Are they different? How so?

We'll revisit the concept of trust in different contexts throughout the book.

2.7 Technical Leader Thinking Limits You to a Tactical Contributor with a Team

What does it mean to be a business executive, as distinct from a cybersecurity functional leader?

As cybersecurity professionals, we generally have to specialize to get started and move up that first rank.

The area in which we developed deep expertise can be seen as the vertical bar of a "T" on the left in the drawing. At the same time, we were generalists in cybersecurity, represented by the horizontal bar. We perform better in our specialized roles when we understand the broader field by becoming familiar with other areas of cybersecurity, often by mere exposure—and, if we want to move up in management, then through intentional effort. We cross over into topics like network security and operating system controls.

When we do this, our area of technical depth provides thinking models and philosophical foundations for understanding cybersecurity as a whole. Then, as our broad knowledge across cybersecurity increases, it informs and shapes our specialized work.

Eventually you see yourself as part of something larger than your domain specialty. This duality of technical skills and broad cyber knowledge serves us well for nearly every position in cybersecurity—until we reach the top position: the CISO.

Then we assume the position of cybersecurity in the "T" on the right of the drawing. The only part of the "T" that remains fully in cyber is the vertical bar. The business executive elements now come in over the top to become the new horizontal of the "T."

Referring back to the stair-step model, you face a similar challenge at each step, and the new open horizontal on the right represents the "next level" activities that you must figure out. The facets to consider during these transitions are knowledge, skills, mindsets, and identity.

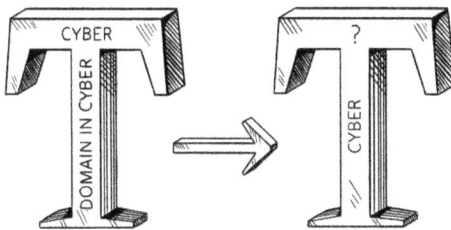

The "T": Two "T"-shaped graphics with an arrow pointing from the left "T" to the right "T." In the "T" on the left, the vertical portion is labeled "domain in cyber," and the horizontal is labeled "cyber." In the "T" on the right, the vertical bar is labeled "cyber" and the horizontal bar is labeled with a question mark. The image represents a transformation from technical leader to business executive.

Making this shift is a linchpin for many CISO goals: getting a seat at the table; being treated as an equal to other executives in the extended C-suite; influencing decisions around mergers and acquisitions (M&A), or at least being consulted early; having "executive presence"; having the "it" factor; and on and on.

Failing to recognize this change and act on it boldly is one of the biggest mistakes that CISOs of all experience levels make.

My executive coaching clients and research participants alike—from the $500 million fast-growth start-ups to the Fortune 100—have made some level of shift to the new "T" to get where they are, but very few make the transition quickly, with clarity, and with confidence, often owing to not having the conceptualization or not receiving support and guidance that resonates.

Specifically, non-CISO executives make assumptions about what other executives know and about how executives think, but these assumptions are not always true of CISOs or technical executives. The non-CISO executives are often flummoxed and frustrated when they try to help a CISO make the shift. They have years or even decades of direct business experience, giving them tacit knowledge that enables them to operate effectively as executives. However, this knowledge is often difficult to consciously articulate, let alone explain to a CISO.

This puts CISOs who came up through technical ranks in an unusually rough spot: misunderstood and difficult for seasoned non-technical peer executives to help.

Questions

1. What assumptions are you making about the nature of the cybersecurity function?
2. What signals may indicate that one of your assumptions needs closer scrutiny?
3. Who or what jumps out at you to help fill in the gaps of being a business executive?

CHAPTER 3:

Mindsets Set You Apart

Getting to space relies on Newton's simple rules of motion—like pushing off a swing, just scaled up with rockets. But navigating your way through town requires GPS, which depends on Einstein's mind-bending ideas about time and space. It feels counterintuitive, considering how daunting one task seems (space travel) compared to the routine nature of the other (finding the nearest coffee shop).

But this counterintuitive applicability of models arises everywhere. In business, the more complicated the model, the more likely it is that someone missed something simple.

Transitioning from a technical mindset to a business executive mindset often means challenging assumptions about your role—and reframing them to uncover unexpected advantages.

"We hire nurses because they know health care."

Early in my career as a CISO for that health-care insurance company—a true nonprofit focused on community benefits—my perspective on organizational culture and excellence shifted.

Unlike typical insurers, the company chose to employ nurses as claim adjudicators, believing their understanding of patient care would benefit members and the company. This approach did have many advantages: greater efficiency, improved care and experiences for members, and a culture that valued care over cost.

An unexpected outcome of the nurse-hiring practice was that 80 percent of the company's workforce were women. This created an environment different from the male-dominated cultures I'd worked in before—and since. Meetings, politics, and budgeting remained, but collaboration flowed differently, intuitively. Problems were solved in different ways. The difference was hard to quantify, but it was visceral; the workplace felt dynamic, supportive, and it was driven by a nuanced, shared commitment to patient care.

Specifically, the company treated the challenges it faced as community health problems, not insurance problems. That difference led to a practically unassailable competitive advantage.

You can seek similar advantages as a leader by challenging assumptions about cybersecurity and redefining the CISO role in a larger frame than what the title would seem to imply—just like those "claim adjudicators" did.

"One hundred and four full-time people doing nothing but clicking a mouse."

In 2009, I was a regulatory consulting leader at a Big 4 consulting firm, working with a major client generating $42 billion in annual revenue. The client had spent a year designing a database to track every vulnerability—potentially thousands—on every asset, assigning a three-level impact to each combination. A full-time employee was dedicated to developing the data model and creating the tool, with additional part-time support from several others. Without automation, the plan was to manually update all assessments weekly to achieve near "real-time" risk data.

I did some quick math based on their estimates of the current environment. To be "real-time," they would need to make fifteen million updates. Per week!

Assuming one second for each update, just for the mouse-clicking, and with no research to see whether anything was fixed or had changed, that's 4,167 hours per week, or the equivalent of 104 full-time people doing nothing but clicking a mouse.

Setting aside the flaws in the model itself, the process was far from practical. Someone had become overly attached to a single approach, wasting more than a year on a direction that was clearly unworkable from the very start.

Had the sponsor had other models, such as vulnerability causality, or asset-to-process mapping, they may have been able to use the other options to see the impracticality of the model they had picked.

This is a clear example of a cybersecurity "leader" not understanding their role as a business executive.

This failure happened because the sponsor was attached to a model that was flawed and disconnected from business needs. As a CISO, broadening your mental models helps you avoid similar pitfalls so you can design cybersecurity solutions that prioritize business outcomes.

3.1 Make the Business Executive Transition to Get a Seat at the Table

If the person building the claims adjudication system had been an experienced accident insurance adjuster who failed to ground themselves in the company's goals, a critical part of the business might have had to have been rebuilt to meet company goals. Fortunately, that wasn't the case.

If the person designing the real-time risk system had paused to consider what the business actually needed—and the effort it would take to achieve it—the proposed solution wouldn't have made it past ideation. It would have saved the company at least a quarter of a million dollars. Unfortunately, that wasn't the case.

In both instances, the lack of thinking in terms of the goals of the business was the difference between relevance and nuisance, business impact and self-interest, trust and mere toleration. Other executives intuitively understand the importance of these mindsets. If you don't, then you won't—and shouldn't—get a seat at the table.

Going back to the "T" metaphor, what questions do you need to answer to define the parameters of the new horizontal during your transition? How do you determine what fills the horizontal bar once you're a CISO?

If you're already in that top position, the good news is that your leaders saw the potential for you to figure it out. Your task is to accept that implicit challenge and ask for their support—support that they are undoubtedly ready to offer to you in some form.

On the other side of this same coin, they are likely waiting for you to contribute to the larger business outside of cyber, using your cyber-honed experience, perspectives, and skills.

Take them up on this by starting simple. Be fully present at meetings for which cyber is only one agenda item, or, when invited, be fully present at meetings for which cyber isn't even on the agenda. Show interest in the business as a whole, not just your corner of it. Pay attention, research what's been talked about, ask questions, and contribute your thinking process even when you don't have the expertise of others. *Apply your heuristics, inquisitiveness, and share your non-cyber concerns. Help others see their ideas in a clearer light through well-considered questions. Use all the traits and skills that make you a good CISO and apply them to the business.*

Before we move on, here's a question—what I call an "unlock" question—that's worth taking a few minutes to reflect on:

Are *you* ready to give up your identity as a technical leader?

> **Note:** Coaching moment. Clients often experience discomfort in the coaching process, even in a safe, supportive space. Productively creating that discomfort is one of the great benefits of coaching. In a book, I can go only so far. So I'll ask: What does it mean to you that I'm asking the question: Are *you* ready to give up your identity as a technical leader? How do you feel about yourself as you're facing this type of explicit challenge question, maybe for the first time?

What if I told you that holding on to your identity as a technical leader is potentially the very thing holding you back—keeping you from your dream workplace situation, from having less stress, from gaining more recognition, or from whatever goals you have and from what you want out of life?

Throughout my conversations with CISOs as an executive coach, in 360-degree feedback interviews with their leaders and peers, in my social research on CISO behavior, and in decades of being a CISO, actively observing CISOs, and working with CISOs, one truth stands out:

Aside from starting a business or facing long-term unemployment, the most challenging career transition that cybersecurity professionals will make is the shift from cyber leader to business executive.

Making the executive shift requires leaving behind parts of the identity that likely once (and possibly still) strongly defined you to yourself. It is similar to the shift from an individual contributor to a manager ("What's my value if I'm not 'doing actual work'?"), but usually with a much deeper identity attachment, and therefore more challenging and complex to navigate.

Making this shift enables you to fill the horizontal part of the "T" by doing something powerful: broadening your perspectives and building the adaptability needed to lead at the executive level. If you're ready, embracing this identity change could be the key to realizing your full potential as a CISO—and perhaps beyond.

Let's overcome that challenge together by reimagining and diverging from the default path.

3.2 Building a Department Helps, but Building Others' Capability Lasts

As a cybersecurity management consultant advising cybersecurity executives, their leaders, and boards of directors, I was often asked for benchmarking.

The most interesting aspects of this category of question are the underlying motivations for asking for benchmarks. People generally wanted to know the following:

1. Are we doing the right thing?
2. Could we be seen as doing the wrong thing?
3. How do we get to the right level for cyber?
4. I have no other means to determine or demonstrate whether we're doing the right things. Can I get benchmarking to give me comfort?

The most frequently asked proxy question for these concerns was about the ideal size of the cybersecurity department.

From my time as a university adjunct instructor, this style of question strikes me as analogous to me saying, "Ah, six pages, it passes the weight test. At least a B!"

At best, this framing and the "answer" is useful as a launching place for a deeper discussion. Importantly, dwelling on this category of question leads everyone into a dead end that's more emotional salve than an answer or indicative of a company-appropriate decision or direction.

The real work is in determining what makes sense now, for this company, in its given situation.

> **Note:** The question itself has a fundamental flaw. What counts? Which parts of the organization and what budget? Do firewalls and their administrators count as a part of cybersecurity or networking? Is identity and access management part of business operations, just like issuing a badge and laptop, or is it squarely a cybersecurity function? Do you count your logging system? Is it pure-play cybersecurity? Or are IT operations using the same platform and they're 90 percent of the usage? The answers create a huge swing in numbers.

So I'd dutifully explain that the question is fundamentally flawed (see note above), to which they never failed to follow up with "Okay, thanks, but what are the numbers?"

After fielding this question for years, I finally had an answer.

"Zero," I'd say.

I'd get smiles and smirks. As I stood there, not adding anything, the range of visible reactions expanded: 10 percent of senior management or the board got it, a quarter froze in fear, and the remaining were confused.

The core idea behind my answer is that if everybody in the organization is operating well, in terms of cybersecurity, then there is little that cybersecurity as a department or stand-alone function would need to do. *The right size of the cybersecurity department is largely based on how much energy and attention is required to get the rest of the company to do what the company would naturally do if they fully understood the risks.*

I always got strong responses to the "zero" answer. Some executives and board members appreciated it because something that had previously eluded them finally clicked for them about the dynamic of cybersecurity. Some people hated it because they took it *only* as the callout that it indirectly was: You have real problems, so stop applying bandages and solve this for real.

> **Note:** As I stood there in front of these groups, I wondered how each person took care of their car. Did they regularly fill up with gas, only to not change their oil unless the repair shop sent them a reminder?

You're better off starting by assuming a zero-size cyber department and building into the business from there.

While this zero-size security department concept was admittedly snarky, it's also a solid thought experiment for designing cybersecurity capabilities—organizationally.

This is another example of the mindsets—one that propels you forward.

Assume a zero-size cybersecurity department. What unexpected, new, and scary opportunities does that create? What solutions does it suggest? Lean into that. Stay with it for as long as you are uncomfortable.

Note: Strongly dismissive of the "zero" concept? As a coach, I rarely call out a behavior directly, but I will if that's what helps that client in that moment. This is potentially one of those moments. If we were sitting in a coaching session and you dismissed the question's concept out of hand, I would ask variations of "What's important for you about this?"

Now imagine everybody in the organization is writing secure code, deploying network equipment in a secure way, and maintaining systems in a secure way, very aware of what's at stake and where the threats come from and how capable those threat actors are.

Wouldn't that be awesome? If so, the "zero"-size cybersecurity department is a starting place, and I encourage you to explore it deeply enough to uncover your assumptions. Because, frankly, we're not going to solve cyber by doing it the same way the field has for the last twenty years.

And to directly address the question at the top: Forget the benchmarking. Seek to structure your department and your rhetoric to get as close to the "zero ideal" as you can, and then ask for what you need to move the company to where it needs to be relative to risk and the fact that the ideal is not the on-the-ground truth.

3.3 Broaden Your Mental Models to Expand Your Value

Mental models shape the way you think about and approach situations—both problems and opportunities. They shape what you see and miss. They have a significant influence on the decisions you make, and even the things you see as decisions. They expand or limit your options. Models also determine the set of criteria you implicitly use to measure your success, or estimate the likelihood of it.

Build a Robust Set of Mental Models to Give You Better Choices and Clearer Thinking

The hierarchy of plants and animals is a wonderful, complete, and useful model.

It's also terrible at helping you understand what makes a thoughtful and appropriate bouquet of flowers for your aunt's birthday versus her funeral.

If you start with the wrong model, you'll never meet others' expectations or relevant goals.

So it goes with all models. Choosing the right model requires the following:

1. Understanding the business context of the situation, question, or decision.

2. Framing it in the form of an addressable challenge or decision
   in that context.
3. Having awareness of *multiple* models relevant to the framing.
4. Selecting one or two models that are
   a. likely to be accepted by others.
   b. useful to you.

It's actually quite hard, but that's it.

Also, not everything needs to be modeled.

Time for some directness. Statistically speaking, most of you will say that you align to the business. It's likely that your process for doing this is accidental, cyber-focused, and slipshod (with all due apologies to those of you who are doing this well).

We can turn the above steps into a diagnostic to get clues as to why you might not actually be aligned.

If you can't get past step 4, chances are you have a situation that shouldn't be modeled for external stakeholders.

If you can't get past step 3, chances are you have a situation where you are not ready or equipped to model.

If you can't get past step 2, you are likely emotionally stuck.

If you can't get past step 1, you lack business acumen.

Avoid Justifying with Math or Points of View That No One Else in the Business Uses

As long as I'm calling out the mistakes bad models lead to, let me be explicit:

1. Don't rely on risk registers disconnected from business opportunities; don't prioritize risk versus risk; don't prioritize cyber risk in isolation from business opportunity.

2. Don't assume that executives share your risk lens, or that yours is relevant to their concerns.

3. Don't treat risk as a separate conversation from growth; don't ignore the opportunity cost of risk treatments.

4. Don't ignore the business's operating models or imperatives in favor of cybersecurity-focused models; don't use heat maps.

> **Note:** Step 1 is a pet peeve that grows into significant frustration for many non-cyber leaders: When was the last time you listed all your risks in life to manage them one against the other? Never? Take a cue from that and stop asking others to do it. Business environments don't fundamentally change human behavior.

What other models are there?

Business models.

I'm suggesting a mindset shift that goes beyond simply moving from a security-focused perspective to one that integrates with broader business strategies. Consider adopting business models and fitting security within those frameworks.

This shift in thinking is part of the journey for a CISO aiming to become a full-fledged member of the C-suite. It brings you a step closer to achieving that goal—a taller stack of above-level achievements on The Readiness Staircase.

Don't Overextend Your Scope Just to Attempt to Justify Your Core Purpose

"Turn cyber into a revenue function!"

That's a ridiculous idea 99.9 percent of the time. Enough said.

If you think it's a viable idea that will work in your circumstances, I'd be interested in hearing that story. Contact me.

Regularly Identify and Consider Additional (Business) Models

Most CISOs aren't naturally drawn to business models, as these models don't directly address cyber and focus on business areas and ways of thinking that few CISOs are deeply familiar with.

It's human nature to use what we know—but it can also hold us back.

Moving beyond philosophical models into implementable approaches, several business-focused practices are relevant:

- Redefine "your team" from your function to your leader and your leader's directs.
- Calibrate your role to stakeholder expectations.
- Focus on and anchor to your stakeholders' interests.
- Accept alignment as indirect, integrative, collaborative, and not on your terms.

The underlying philosophical perspective to adopt is that cybersecurity lives within broader business goals rather than isolated as a stand-alone function with its own models to be pushed onto the business.

Failure to orient to business models is a lack of thinking broadly and specifically—and a failure that can be avoided.

It bears repeating:

Start with models appropriate to business goals; otherwise you'll hit your cyber goals only by sheer luck or accident, or think you have when you unknowingly have not but do not realize it due to the silence of others.

We will discuss developing business-focused practices and calibration in greater detail in part 2.

Questions

1. What models do you use that feel played out or stale?
2. Who can you learn from about how your organization thinks of itself?
3. What cyber-honed executive skills can you apply outside of cybersecurity?

PART II:

Relationships

CHAPTER 4:

Ground Yourself in the Wider World; Skip the Window Dressing

"I have a calculator."

A calculator being held up by a hand.

In high school, I ran for treasurer during my senior year. Three of us gave speeches: the future class valedictorian, me, and a good friend I'd known since elementary school. The class valedictorian spoke first, then me. We both talked about the budget, fundraising through doughnut sales, and other finance-related topics. My friend stood up and said, "I have a calculator," held up a calculator, said a few words about not making math mistakes, and sat down. The class roared with laughter.

He won, I came in second as deputy treasurer, and the valedictorian finished last. This shouldn't surprise us: People win by talking about relevant things in clear ways. In most circumstances, the simple message is also the most powerful—even, and especially, in weighty situations. In the case of the treasurer role, they thought of it as keeping the books accurately.

Don't say more than you need to. Don't overexplain. Don't count on how smart or right you are. And don't bore people with what you think is important. Understand what they think is important.

If the average CISO had to present to my high school senior class, they'd learn fast what works and what doesn't.

Cybersecurity and control frameworks are often like my treasurer campaign speech: overcomplicating what is, at its core, a much simpler conversation.

Consider how you communicate by default.

In the rest of this chapter, we will discuss the advisors and networks that will help you succeed as an executive. In the following chapter, we will discuss the workplace stakeholders that determine whether you succeeded in your role.

Let's get started with advisors.

4.1 Others' Solutions Often Complicate Your Path

Among CISOs, there's a circulating list of approaches for trying to gain attention and resources from business leaders. Given the limited options, many cybersecurity executives end up reworking variations of these strategies, hoping for real success—or at least the appearance of it.

These techniques can work—but only under highly specific conditions—namely, when the company imperatives, cultures, personalities, individual motivations, and circumstances are conducive. That's a lot to line up, and it doesn't usually happen. Importantly, what worked for one CISO in a particular environment with their unique strengths and styles may not translate well for you, even in similar circumstances.

Despite ample evidence that the often-cited techniques and simple platitudes fail far more often than succeed, it can be hard for CISOs to see this pattern, even when comparing notes with each other. At a certain point, desperation overrides logic, just as with lottery players who say to themselves, "If I can just pick the right numbers one time, I'll be all set." That might provide a glimmer of hope, but it's a terrible way to lead a corporate function.

So basing your choices solely (or largely) on stories from peers or on conference anecdotes is like playing the lottery: You might occasionally win, but that's hardly a sound strategy. Having only one way to do something and hoping it will work is similarly a poor strategy.

The keystone of executive success isn't a specific technique but adaptability—the ability to draw from a repertoire and pick the right approach for the situation. *Without versatility, success is accidental.*

CISO Affinity Groups Help, But Not Usually to the Degree or in Areas You Most Need

Many CISOs I speak with say their primary source of guidance comes from CISO peer networks, even while questioning the value of them. The reality is that CISO networks too often leave a CISO feeling alone once they're back at work. Swapping stories can be helpful, but it rarely makes a significant structural impact.

When you go back to work, to your day-to-day, who's rowing with you?

4.2 A Wide Circle of Advisors Increases Your Value and Creativity

Seeking advice from diverse perspectives often brings unexpected clarity and insights—not just about what to focus on, but also what to let go of or how to connect ideas.

"Atoms vibrate like violin strings."

Legend has it that renowned physicist Niels Bohr drew an analogy between atomic systems and music from a conversation with his musician brother, Harald. Bohr's concept of electrons orbiting in quantized orbits came in part from his brother's explanation of vibrating strings, where each frequency creates a unique pitch.

Whether it's fact or myth, the model of atomic structure, which has analogies to music, was a revolution in physics.

Seeking advice from diverse perspectives often brings unexpected clarity and insights—not just about what to focus on, but also what to let go of or how to connect ideas.

"They solved a longtime challenge in cabled–cantilever design without our expertise."

At Burning Man, I met a German structural engineer who'd been working in commercial architecture for decades. She mentioned that she'd been coming to the festival for years simply because she enjoyed it. But this particular year she'd found something she didn't expect. She explained a persistent problem in anchored cabled-cantilever design—one that experts in her field hadn't cracked after a decade of effort and for which no solution was even on the horizon.

Then she described what she saw: A group of artists, builders, and carpenters at Burning Man had solved it. "These people weren't even professional engineers; they're artists, music lovers, and amateur carpenters," she said, "but they built a working structure as if it were just another sculpture."

The breakthrough didn't happen in a traditional engineering setting. It happened because everything was different there—different people, different goals, and no attachment to conventional ways of thinking.

You don't have to go to Burning Man, but you do have to get outside your in-group if you're looking for fresh ideas and new solutions.

What wide-ranging fields do you want to have in your circle of advisors?

Sourcing New Ideas and Outside Perspectives Drives Soundness of Thinking and Develops Your Range and Adaptability

We are strongest when we collaborate with people who introduce us to new ideas and perspectives. Pick the right feedback provider for your goals at that moment. Choose based on these factors. Optimize for as many as possible based on the situation.

For instance, seek people who are disinterested in the situation, who might not have a stake in the outcome.

Go Broad with Your Personal Circle of Advisors, Because It's Not Who You Might Think

Avoid staying within field-specific dogma, where challenges to assumptions can turn into unproductive debates or unquestioning acceptance.

In contrast, challenging an entire way of thinking can help clarify your ideas and—with enough openness—might even shift your perspective or introduce entirely new sets of tools. While deep insights from within your field are valuable, they become increasingly rare as your experience grows.

Go outside the field. Go broadly and be specific about your goals.

Distinguish Among the Potential Modes of Advisors and Networks

While you may have many advisors, they operate in three distinct modes:

- Development: helps you as a professional (inside and outside your organization).
- Calibration: helps you define your role (inside your organization).
- Organizational: helps your role achieve business goals (inside your organization).

One person can often operate in more than one mode. For

example, an internal HR advisor might help with both development and calibration.

The possibilities are endless, but to make the most of your time with an advisor and of their time with you, understand your needs, what each person can truly offer, and whom to seek out.

In this chapter, we focus on development mode. In chapter 5, we discuss how to work inside your organization by leveraging all three modes.

Your Circle of Advisors for Development Mode

Typically, five to eight people is a good starting point for your inner circle of development-mode advisors. Fewer might lack diversity or perspective, while more makes it hard to maintain meaningful relationships. Your inner circle will evolve naturally, and you'll rotate advisors as relationships run their course or new perspectives are needed. You engage the rest of your advisors to maintain relationships and call on them as needed.

4.3 Advisors Fill In What You Lack

For a CISO, building a broad set of advisors accelerates professional growth and leadership effectiveness. These networks provide opportunities to learn, gain insights, and refine your leadership skills by tapping into the perspectives of those outside the cybersecurity field or from a distance.

Types of Advisors

Advisors can be either stakeholders or non-stakeholders. Non-stakeholder advisors offer advantages over stakeholders.

Consider developing relationships across the following types:

- Internal Coaches: People within your organization who can help you understand the organization, but who are distant enough from what you do that they have little to no direct stake or organizational interest in your decisions.
- Functional Guides: Deep experts with both knowledge and cultural understanding of a specific function who can offer insights into how that function operates.
- Cross-Domain Executives: Leaders from other business areas whose perspectives on risk, opportunity, and strategy can broaden your thinking.
- Professional Mentors: Established professionals in the cybersecurity field or beyond, either within your organization if they don't have a significant stake (a VP in another business unit, for example) or from outside your company.
- Professional Coaches: Professional coaches who specialize in executive growth:
  o Executive Coaches: Help with overall leadership development, communication style, and self-awareness.
  o Leadership Coaches: Focus on building specific leadership skills, such as strategic influence, stakeholder engagement, and decision-making.
  o Career Coaches: Guide long-term career planning and help you navigate C-suite aspirations or transitions between executive roles.
  o Specialty Coaches: Include communication or brand coaches, who can be particularly helpful if you're preparing for public speaking or enhancing your professional visibility.
- Peer Groups: Share experiences and insights (if you have not already, you can join or form peer groups with other CISOs).

Guides Who Understand the Company and Its Culture

Consider finding an inside guide—someone who knows the organization deeply but who's not in your chain of command or directly influenced by your decisions or success.

These guides are a resource for understanding your conversations with the people you need to connect with. If you're fortunate, you might include the former CISO among your advisors—though not exclusively, as they bring their own biases, and this can also be politically fraught.

Breakthroughs often emerge where fields converge, reshaping how we approach challenges. For cybersecurity leaders, stepping beyond their domain and embracing cross-disciplinary thinking is critical to fostering innovation and staying relevant.

Include people across layers—bosses' direct reports, veteran employees with institutional knowledge, middle managers with operational insight, and key project leads driving initiatives.

Resources and Associations

In appendix E you'll find a curated list of professional associations and organizations that may be helpful to follow or join. These groups can expand your circle and help you stay informed about industry trends and leadership practices.

4.4 Overreliance on Certain Advisors Invites Disaster

How does one pick advisors?

Advisors can come from your list of former bosses and colleagues. More generally, this list could, and probably should, include former or stakeholder-distant current colleagues from human resources, legal, internal audit, recruiting, sales, finance, marketing, and operations.

Cybersecurity executive colleagues tend to accrue other cybersecurity executives by default, but if you don't have at least a few of those connections, you should develop them.

As you contemplate individuals, consider who can both inspire and challenge you to think in new ways and help you recognize areas for improvement. The candidates shouldn't be overly attached or dependent on the current you.

> **Note:** Be circumspect about picking friends as advisors. They don't have exposure to you in the workplace, thus limiting and potentially skewing their insights about you as an executive. They are often also attached to the person they view you as today, or even yesterday.

Due to my personal style and approach to organizational dynamics over my history as a cybersecurity executive, my metaphorical Rolodex contains several people in each of the fields I suggested for you, as well as others. More recently, as a coach, writer, speaker, and business owner, I've added business owners and coaches who specialize in various types of executive coaching, leadership coaching, and business coaching. My list also includes artists, teachers, and others. It includes far too few local EDM producers and DJs, but that's a different story.

Good advisors are rare. Cultivate them. You may not initially know what makes a good advisor for you, and when you should have one, but keep practicing until the distinction becomes clear.

4.5 Curiosity Opens Doors; Personal Interest Keeps Them Open

Approaching potential advisors might be as simple as saying, "Hey,

I trust what you say. Whenever we have a conversation, I walk away better for it. You make me think, without telling me what to do. I'd like to meet, if we could, once in a while." Or "I would really love to just set aside some time and talk occasionally. I think it's going to help me."

People are often very willing to do that. I've never had anybody say no without offering sound advice or a good direction. People want to help, and people feel good helping. And done right, the person who's helping you grows through the process, too, and gets something from it themselves.

Cadence Is Mutually Decided, So Don't Force It

Establishing a cadence is up to you and each advisor. Outside of a paid professional advisor, such as an executive coach, more than once a month is almost certainly too often. Once a year might not be enough. The cadence doesn't have to be regular; it doesn't have to be rigorous. But it should be somebody that you plan to connect with on more than just an as-needed basis. Having a lifeline is helpful, but more growth happens when you set aside time for it.

Who Not to Pick Can Be as Important as Who to Pick

For your professional circle of advisors, leave off most friends and family. ALWAYS skip those who doubt you. Don't consider cynical people, energy vampires, those who create drama, people who don't bathe as an act of rebellion, naysayers, and anybody who isn't in it for you. Yes, there's an easter egg there—a reminder to decide for yourself. Maybe non-bathers are okay with you.

Structure

I'm calling them a circle of advisors, but I meet them separately.

Together would be interesting as a dinner party, but probably not for what you're trying to achieve.

Me as an Example

Using myself as an example, I have ten advisors at the moment.

My four formal advisors include my executive coach, video branding coach, book author coach, and book marketing coach.

My professional peer advisors as a coach include those with backgrounds as a university president, industrial psychologist, human resources executive, and sales executive.

I speak with other CEOs, and of those, about half are in or adjacent to cybersecurity. The other half are entirely unrelated to cybersecurity.

In short, only some are cybersecurity. This has been my practice for twenty years. The point isn't to surround myself with people who think like me (CISOs, or even now, executive coaches); the point is to surround myself with people who have a point of view that's different from mine.

I seek people who have seen and dealt with a lot, but who don't talk about it much.

What style do you want in your circle of advisors?

Note 1: As part of my research and method development, and to stay tuned in to the specific challenges my typical but noncurrent clients face as well as the general zeitgeist, I speak with a dozen CISOs and their executives about the CISO role each month. To gain fresh perspectives on how CISOs approach their challenges, I have about five first-time conversations of thirty to ninety minutes. These leaders come from a range of organizations, from small and midsize businesses to large corporations, nonprofits, government, and the military.

Note 2: As a business owner, I don't count advisors such as my adept executive assistant, or my lawyer, accountant, or bookkeeper. Nor do I count my past marketing strategists, lead-generation consultants, or two public relations firms.

The "T," Part 2—Identifying the New Horizontal of the "T"

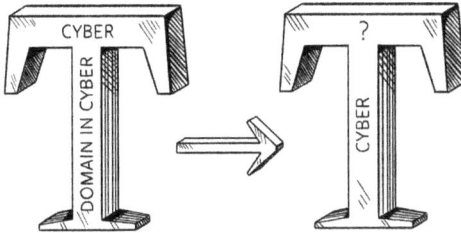

Colleagues will continue to recognize your technological expertise, a core strength you bring to the business, but consider focusing your selection of advisors on one thing: understanding what goes into the horizontal of the "T" on the right, and filling it in.

Mixing In More Formal and Paid Professionals or Paid Groups

Beyond advisors, several other structures and arrangements can serve you well. At the simple and low-cost end, your local Toastmasters advanced club, religious or spiritual community advisors, and peers on nonprofit boards can be potential candidates for your advisor group.

Mentors

Many people want to mentor or coach, or say they do. Most just have a chat with you. Some are good. Few are outstanding. Even fewer regard mentoring seriously enough to view it as a distinct skill or practice to develop.

Masterminds

The word *mastermind* is bandied about, often used to describe groups that are not masterminds. If the group spends much of its time listening to presentations and talking heads, it's not a mastermind. If it's about technology solutions but claims to be for executives, it's not a mastermind. If you don't have to go through a review for values fit and you're allowed to join the group without meeting other members to ensure a good match, it's not a mastermind group.

In general, be wary in all these cases so you can avoid being a member of a captured group to be sold to, when what you really need are leadership and executive sounding boards.

To be clear, well-crafted and meticulously and insightfully designed masterminds will be solely for the members' benefit and will be a professionally facilitated discussion. The point of a mastermind is for members to talk with one another in highly useful and productive ways. The leader of the mastermind should have a solid background in structured dialogue, facilitative techniques, and insight generation. An effective mastermind leader will have emotional intelligence and the self-management skills to lead without injecting too much of themselves.

There are great masterminds, but they are few and far between. Good ones will be expensive. You can expect to pay between $10,000 and $50,000 per year.

Paid Advisors

As mentioned, I have professional (paid) advisors in my circle. Although they can be expensive, professionals also offer significant advantages.

First, this is the easiest way to have someone commit to you. It's also a way to get outside of your circle.

Second, if you plan to be on the speaking circuit, write a book, or become a Fortune 500 CISO, you will benefit from paying exceptional special advisers to help you get there. And fair warning, that is a significant investment of time, money, and brain cycles.

At the center, an executive coach is a key to many major goals. A speaking coach can help you polish, a book coach can help you find your writing voice, and a brand coach can help you develop your public persona, *but an executive coach helps you find you.*

Questions

1. When you last sought advice, did it genuinely challenge your assumptions or reinforce familiar ideas? How might you find more challenging perspectives?
2. Recall a time an outsider's perspective helped you solve a problem unexpectedly. How can you invite similar insights into current challenges?
3. Do you have sufficient advisor support in "development," "calibration," and "organizational" areas? Which type do you rely on most, and which could you seek more actively?

CHAPTER 5:

Business Influence Develops from Genuine Interest in Others' Concerns and Goals

"The Wrongest Layoff"

Shortly into my first global management position, my director informed me layoffs were imminent and decisions had already been made for my team.

I was told to sit across from real people—recently my peers—and tell them they no longer had a job.

These were friends and colleagues, some of whom were in the midst of big life events, and all of whom I cared about. One had just gotten married; one had just bought a house. It was awful.

But one hit me differently. Even though I couldn't fully articulate it at the time, I knew it was just "wrong." The person was a dedicated, high-performing employee, often working nights and weekends due to the demands of his role. Yet the primary complaint against him was that he occasionally arrived after 9:00 a.m.—for legitimate health reasons. It was fundamentally unfair.

I tried pleading with the appropriate person, but I also knew this person had significantly contributed to the decision in the first place. At the time, I was not seasoned in informal organizational influence. I hadn't considered this scenario of fighting for someone during a layoff. I hadn't spent much time on anything like it.

I had been wrapped up in tech and operations, largely on what was under my direct control.

That was my job, after all, wasn't it?

In the end, I sat across the conference table on the designated day and laid the person off.

For him, it was manifestly unfair. For me, it was one of the lowest and most disempowering moments of my career: a near-total lack of any sense of being able to make the organization do the right thing.

I think he took it better than I did, honestly.

Over time I realized that my lack of exploration and practice with scenario planning made me doubt myself at every level. I hadn't done the management equivalent of running plays in my head.

I unknowingly followed the default option presented to me. I accepted it, incorrectly, when faced with my director and an HR professional who said, "You have to do it." Surrounded by others who seemed so certain, I began to doubt my own stance and assumed that the "experts" must be right.

I was arguing with the wrong people and on the wrong basis, because I didn't know how to use the right connections. I couldn't see other choices. I hadn't run the plays in my head, or with anyone else. I defaulted to solving problems on my own.

I'm going to acknowledge to myself that this was early in my management career, but I will also offer some advice:

- Think deeply about what is important to you personally (ethics, values, principles).
- Consider the scenarios ("good" and "bad") you can foresee facing.
- Identify several ways to develop options—beyond those that are presented.
- Run plays. What will you flex? What are your nonnegotiable ethics, values, and principles?

When you eventually run into your version of "the wrongest layoff," you'll be primed and ready to:

- Get perspective from your advisors on those options.
- Work hard to find more options until you have several palatable options.
- Leverage your relationships to make your case effectively.

The me of twenty-five years ago is in many ways a stranger to the me of today. While I can't know the outcomes had I followed the above practice, I would have been truer to what I think is important. Applying what I knew about myself to potential situations would have helped me understand myself better and prepared me more for the situation. In the end, I might not have been able to change the outcome, but personal clarity and earlier practice of building professional and workplace support would have helped me advocate more effectively for this person.

I committed to myself that I would take the logic of doing DR and incident response (IR) scenarios and apply them not just to technical issues, but to a range of circumstances. I would use the scenario-planning muscle in new ways. As Dwight D. Eisenhower was reputed to have said, "Plans are worthless, but planning is everything."

Missing small issues is small. Missing the opportunity to practice for large issues can be catastrophic.

"Let's have you talk to some people."

Years later, in the first CISO role that I mentioned earlier, my CIO—without my realizing it—nudged me into regular meetings with leaders across the organization: the chief strategy officer, the chief operations officer, the chief financial officer, the corporate counsel, the attorney responsible for interpreting HIPAA, the VP of HR, the director of internal audit.

My CIO prepared me so well I barely noticed I was meeting regularly with the most-senior executives across the company—a rare privilege, then and sometimes even now, for CISOs—and I learned a lot.

The relationships I developed introduced me to a new way of thinking about the business and the relation of my work to it. Cross-functional engagement cultivated my business acumen in deeper ways than an MBA could have. We found problems together, and solved them together, and the collaborations gave birth to new ideas grounded in the business. I developed new approaches and perspectives from the process itself, and I use them to this day.

Imagine if other organizations embodied this kind of approach with their CISOs. Imagine if you did this for your direct reports.

Lacking that, imagine what it would be like for you to develop approaches and tools to do this yourself.

5.1 Business Acumen Is a Prerequisite for . . . Everything

I often hear, "We're aligning cybersecurity to the business!"

Oh, really? That's nice. But my follow-up question often reveals the reality: "How are you actually doing that?"

Ouch.

This might sting, but what I'm compelled to say is that in far too many cases the CISO's implementation of "alignment" is ad hoc, accidental, and abysmal.

What "aligning" often boils down to is the CISO expecting non-cybersecurity executives to learn about cybersecurity or the projection of a "culture of cybersecurity"—one that's somehow bolted onto the organization's true culture Frankenstein-style.

That's not alignment. That's self-delusion.

If the CISO doesn't know significantly more about a stakeholder's function than the stakeholder knows about cybersecurity, or if the CISO doesn't have business acumen and situational understanding, there's not much alignment.

Evaluating Your Commitment to Business Alignment

How do we move from ad hoc, accidental, and abysmal to effective?

For one, we can measure the inputs—or steps—that help us get there. For CISOs, a significant factor is *business acumen*. Start by evaluating how much effort you put into understanding the business, compared to how much you expect others to understand cybersecurity.

When setting your goal, aim to exceed—not just match—what you expect other executives to understand about cyber. There are two reasons for this:

First, the business isn't about cybersecurity. Cybersecurity exists to support the business. Revenue preservers support revenue generators.

Second, the Dunning–Kruger effect is at play. What you know about your field that others don't vastly outweighs what you know about theirs. It's a trap. The simplest way to overcome it is to accept that bias is present and deliberately build in a counterweight—probably a big one—to your efforts and assumptions.

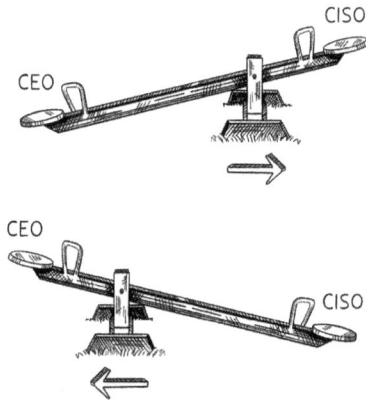

As the CISO forces executives to operate in a cyber mode (fulcrum closer to cyber, top) the results are unpredictable and rare, even if big. As the CISO moves towards the business (fulcrum closer to the business, bottom), the wins are smaller but more certain.

> **Note:** Business acumen is highly susceptible to the Dunning–Kruger effect, so you'll need to find objective units to measure this, such as learning hours per year. The Dunning–Kruger effect is a cognitive bias where individuals with limited exposure in a specific area overestimate their competence, while those with deep expertise may underestimate their abilities due to a heightened awareness of the field's complexity.

Areas in Which to Develop Business Acumen
You can learn a lot about business, from the global to the specific-ish:

1. Strategy: Understand strategy models, business models, when to use each, and which are relevant to your organization.

2. Industry: Familiarize yourself with industry profit levers and key assumptions. For example, in nonluxury retail, tight margins mean that brief operational delays can lead to financial losses, while in B2B consulting, this is rarely the case.

3. Company Market Strategy: Know your organization's unique value proposition, competitive strengths, and market approach—whether it's low-cost, first-mover, white-glove service, or another model.

4. Company Value Creation & Chains: Identify core activities that generate value, from product development to customer support, and understand departmental interdependencies.

5. Company Culture & Practices: Learn the formal policies as well as the informal norms that shape daily interactions, conflict management, and innovation within your organization.

6. Company Power Structures: Recognize both formal processes (like annual budget cycles) and informal ones, which reveal who truly influences decisions and how.

7. Executive Microculture & Practices: Dive into how executives collaborate, communicate, and make decisions. These "microcultures" often differ from the broader company culture and play a critical role in your success as an executive.

8. Professional Ethos & Values: Each profession operates within a general ethos and logical framework that influences decision-making and behavior.

While development-mode advisors can directly support 1–3, and may offer insights for 6 and 7, you'll need calibration and organizational mode advisors for 4–7, and to conduct your own research of professional cannon and codes of conduct for 8.

Evaluate the Maturity of Your Orientation to the Business and Your Executive Stance

Traditional cybersecurity maturity models assess program effectiveness, focusing on where a cybersecurity program stands. But business executive maturity demands a different approach—one that prioritizes personal growth, adaptability, and orientation to executive team play.

In a departure from program-based models, I designed the first model of its kind: the Cyber-Business Orientation Maturity Scale™ (CBOMS), which specifically assesses the CISO's performance as a business executive.

To support self-development from reading this book, I've adapted CBOMS from an evaluative tool—measuring current or historical performance—into a formative guide designed to foster understanding that leads to continuous executive growth. The version presented here will help you assess your alignment and clarify where you can and want to go.

MATURITY	EXECUTIVE STANCE	INDICATORS
1. Crawl	Technical Specialist	• I am a technical expert and know what cybersecurity needs to do. • I benchmark myself against the technical skills and knowledge of other cybersecurity executives. • I focus on established tenets of cybersecurity, and these guide my work and priorities.
2. Walk	Business-Aligned Expert	• I use my technical expertise to align cybersecurity with business needs. • I mainly learn about business plans from my direct manager, and I sometimes request general updates. • I ask business stakeholders what they need from cybersecurity.
3. Run	Business-Engaged Strategist	• I set cybersecurity focus areas based on company goals and regularly refresh my knowledge of our industry. • I make it a point to understand each executive's functional area and role in achieving our business objectives. • I actively solicit business plans and priorities from non-cyber executive peers.
4. Fly	Business-Centric Collaborator	• I collaborate with peers on company-wide challenges and prepare cybersecurity to support future business directions. Peers keep me informed of future opportunities and challenges. • I am invited to contribute to business decisions beyond cybersecurity, bringing my unique perspectives. I may be asked to take on additional roles outside of cybersecurity.
5. Soar	Integrated Business Leader	• My role as CISO is fully integrated with C-suite executive peers, with cybersecurity operations run by my direct reports. • I am at the table for business planning and strategy discussions that involve more than a handful of the most-senior leadership team. • I am seen as an equal contributor to business growth and problem-solving. • My name might even come up for consideration in CEO succession discussions.

Note: You might be asking yourself, "What about everything I know about department planning, budgeting, resource management, team development, project oversight, and other general management skills?" If you believe that any of these things are your edge, you're not only not on the playing field of executive leadership, you're not even in the parking lot of the stadium.

Now take a moment to consider where you want to be, and by when. Be bold but also fair to yourself.

Who do you need to become to achieve your goals?

Now that you have given that thought, let's talk about assumptions that stand in the way.

5.2 Technical Expertise Is Only a Small Fraction of Professional Trust

Upon reaching the executive level, you're expected to make a shift, because the people you work most closely with are different, and they operate differently.

To cite noted business consultant and executive coach Marshall Goldsmith: "What got you here won't get you there."

It is rare to see a CISO who takes this far enough. CISOs almost always experience hesitation or drag when leaving the familiar realm of cybersecurity, program management, and functional area leadership.

But there is a way out, a path to being a member of the CEO's team, and this path addresses many of the challenges cited by our cybersecurity executive peers today.

Making this shift also serves to protect you in tough situations.

Let's start with some foundations.

The Trust Equation

Building trust is one of the core capabilities of any leader.

But while we all have a sense of when trust is present or when it isn't, we don't often consider what actually goes into trust and who we need to build it with.

Unexamined, trust seems like a feeling—something unpredictable, formless, and out of our control.

What if there was a model for trust that we could use to understand why people trust us as leaders, and how we can engender genuine trust through genuine behaviors, simply by understanding what goes into trust?

David Maister, author of *The Trusted Advisor*, developed such an equation.

Trust is the sum of credibility, reliability, and intimacy, divided by self-orientation:

- Credibility—Expertise and clear communication.
- Reliability—Consistency and follow-through.
- Intimacy—A sense of safety and openness.
- Self-Orientation—The degree to which you focus on yourself rather than others.

Even with strong credibility, reliability, and intimacy, high self-orientation erodes trust. If people sense you're acting in your own interest—rather than with genuine regard for theirs—trust breaks down. Importantly, common interest is not the same as low self-orientation.

You can share a goal with someone yet still be perceived as self-serving if you frame it only through your own perspective, rather than considering their needs, concerns, and priorities.

Some definitions:

Credibility is that sense your audience has that you know what you are talking about, because you have relevant experience, knowledge, or insight on the subject. It in part comes from observing what we talk about, how we frame it, how we present ourselves, and the way others experience our expertise.

Reliability is actions, our predictability, and whether others find us dependable. Types of reliability include congruence from words to actions, providing support when needed, and offering specific skills in a consistent way.

Intimacy entails our closeness to others, openness about ourselves and our vulnerabilities, and our ability to manage ourselves so that others feel comfortable confiding in us and perceive us as having discretion. Intimacy enables people to treat others and their circumstances with empathy, care, and consideration.

As these three things increase, trust goes up.

But the fastest surefire way to destroy trust is self-orientation—that is, being self-interested or self-concerned. For example, in a conversation about system vulnerabilities, if you focus on yourself and your goals as a leader, you demonstrate that you are not oriented to the other person. You can come across as transactional, tactical, and a host of other "bad" characteristics. Maybe even a career climber or shyster.

Think about a time when you had a visceral (body-felt) sense that you couldn't trust someone.

Was the person you were dealing with clearly interested in themselves, and not at all interested in you? The other person may have given you a signal that you interpreted as "I am willing to harm you or overlook your needs to get what I want."

It's not a very good feeling, and it's one that quickly destroys trust.

As a CISO, how often do you show up as being just about yourself and your cybersecurity goals?

Are you engendering or destroying trust in your everyday behavior?

Where This Shows Up

Many arguments for cybersecurity support, resources, and budget can be reduced to the following:

- Because Cyber (mentioned earlier).
- I'm here to help (support me).
- Ignore this issue at your peril (you're an idiot).
- You need me (you're still an idiot).

If you've never asked what someone cares about (the "patient care" unlock from earlier), your conversations can't be about much else than you. If you don't know what they need as a business leader, or even as a person, how could it be about anything but you?

Messages that embody communication like this—no matter how covered in embellishment—at a certain point are more likely to create a most natural human reaction: polite disregard.

Working against human nature works against you—no matter the nobility of your intent.

What to Do

Listen with genuine curiosity. Start there. If you do this right, your empathy will be nearly automatic. That's a game changer.

Envision what that might look like for you.

Simply: Be curious and put the other person first without putting them on the spot. People appreciate that.

5.3 Relationship–Building Isn't the Same as Repeating Platitudes

Many CISOs succeed and fail based on acting alone *versus* working effectively with peers and leaders. That means having mutually beneficial or satisfying relationships with them.

While there's nothing particularly innovative about having good relationships with your stakeholders, it's also easy to talk in platitudes and not take effective action.

How do you get beyond simple platitudes? For most, it comes down to curating a wide selection of approaches and tools, having the experience or support to select better approaches, and building confidence with practice and experimentation in using the tools.

Consider the following foundational questions before starting a conversation:

What's your objective? Have a few loose—even uncertain—ideas about how you each might benefit. Be clear about your broader goals, whether it's gaining support for cybersecurity initiatives, understanding business priorities, or fostering collaboration. Then set your objectives aside. Knowing what you want to achieve beyond the transactional level helps guide your actions, but focusing on the other person is how you'll truly get there.

What's your approach? Consider a variety of approaches, such as stakeholder-centric or common-purpose discovery, while balancing credibility, reliability, and intimacy against self-orientation. Tailor your approach to fit your style and strengths, and the needs and expectations of each stakeholder, to ensure your interactions are meaningful for them.

What's the ideal cadence? This might include a mix of regular check-ins, quarterly reviews, and ad hoc meetings. Consider the availability and preferences of your stakeholders. Allocate time for both formal meetings and social and informal interactions. The latter two can provide invaluable insights and strengthen relationships in a more relaxed setting.

What's your maintenance? Review and adjust your relationship-building strategies based on feedback and changing circumstances and input. Consider tracking your interactions, key learnings and insights, and thoughts about what worked for you and didn't. Let your conversation partners know what you did with what you gained. Indicate what actions you took, even when those are thought actions such as a decision or new perspective. This confirms your openness to being influenced, part of the implied social contract, and it signals that you are someone worth investing time in.

Networks Thrive on Understanding, One Relationship at a Time

People come to the table with good intent, statistically, and only a small portion pursue gain without regard for others (an estimated 1–3 percent). However, good intent can be lost or misconstrued by assumptions about the personal meaning of business decisions, actions, and behaviors. One way to address potentially damaging assumptions is through relationships.

> **Note:** Whether someone has good or ill intent, or carries personal baggage, is not yours to identify, diagnose, or assume. Don't go there—it's not your place to speculate on someone's history or perceived faults. Such thinking is a largely unhelpful mental distraction that keeps you from focusing on what matters: influencing where you can.

With this premise—what is a relationship?

In the everyday business world, a relationship tends to occur between two people who share values and have overlapping goals or interests. There are other criteria, but these are the main two.

By virtue of working at the same organization, you probably share values. Someone who regularly volunteers to do habitat restoration is probably not employed by a large oil producer, and it's unlikely that a deregulation proponent works at a hospital specializing in asthma and emphysema. So let's assume there are at least some overlapping values in the workplace. That's a starting point.

Discovering what those shared values are and how they are interpreted and acted upon, however, poses unique challenges. First, not all senior executives have thought through their values. Asking for someone's top three values may not provide information that is accurate or substantial enough. Practically speaking, it may force someone to answer before they've had a chance to consider.

That aside, a single word can have multiple implications. For example, a former boss and I both value integrity, but we had different definitions, mine focusing on a congruence between intent and promises (be forthright) and his focusing on a congruence between stated intent and actions (meet commitments). Those can both be representations of integrity. But they come with different expectations and implications.

Well-considered questions about passions, interests, and personally meaningful events offer insights into values: important stories from someone's career and life, people they admire and why, pet peeves.

I will also say this: In the history of cybersecurity (and possibly the world) the most difficult thing is to get someone to intrinsically care about something that they know little about and that doesn't plainly and manifestly connect to an existing goal, passion, or concern.

So these types of conversations—sometimes deemed not "practical" in workplace settings—are actually important, and tend to happen in social situations. This is why lunches, work holiday-time events, and golf can be practical and useful activities, even when they don't seem "productive." To have value, not every work activity has to be focused directly on the business itself. Yes, and I hate to admit it, even golf.

Discovering Goals

How likely is it that your colleagues are motivated by cybersecurity? Sure, they don't want anything to go wrong. But they are probably not motivated by it.

The CNO in the story at the beginning of chapter 2 wasn't—and, arguably, she probably wasn't even after her significant change in stance.

So why do cybersecurity people keep barking up that tree?

Self-Imposed Barriers

The self-imposed barriers that we create include not having direct, personal insight into the following:

- What the other person cares about.
- What the other person is facing—their challenges.
- The other person's goals or aspirations.
- The other person's values and principles.

These gaps are easily fixable with conversations driven by honest curiosity and backed by light but thoughtful planning.

5.4 Hoping to Find Common Ground Is Not the Same as Building Common Ground

Every CISO stakeholder development conversation is predicated on a few fundamental categories of questions. Early stakeholder conversations can and should focus on answering or enhancing existing answers to these questions, and most stakeholder relationships benefit from refining your understanding in every conversation.

There's no room for you to make assumptions about answers to these questions—each assumption becomes a land mine for you to disarm later on.

Core Questions for Executive Stakeholders

When you're a member of the executive team, your conversations should naturally move toward discovering and refining the following:

1. How do executive stakeholders see their roles?
2. How do they see their roles playing a part in the success of the organization?
3. What do they hope for in their roles and careers?
4. What challenges do they face in their roles and careers?

5. In light of this, how do they see the CISO role, and what do they need from it?

Against the backdrop of those questions, how can we have better conversations?

5.5 Conversations Are Exchanges of Ideas, Not a Personal Platform

For a conversation to be considered successful, each party must leave with at least a changed perspective, even if slightly.

If a CISO shows up to a conversation, makes a bunch of points, and doesn't listen enough to be influenced by the other person, that's a failed conversation. That's breaking the social contract around influence.

Getting It "Right from the Start"

Here's a fictional conversation a CISO starts with their boss. For the purposes of this example, I'm making the nature and tone obvious. It's not uncommon for more seasoned leaders to avoid this level of obviousness while still following the same tack.

I invite you to ask whether you are having similar conversations, even if you are more tactful about it.

"I'd like to talk about our disaster recovery plan. [Insert long history, background, or situation explanation.] It hasn't been tested in two years. We need to test it; otherwise it could have a big impact on the company."

Average, typical, terrible.

The CISO didn't invite the boss into the conversation, but rather expected the boss to orient, frame, and jump in, 100 percent on the

CISO's terms. What if the boss wanted to have a slightly different conversation?

Implicitly, the CISO is seeking support—but they haven't negotiated with the other person that this is the conversation that the other person wants or needs.

Frankly, the CISO hasn't earned the right in the conversation to make their demands.

Almost worse, the CISO then subtly but very annoyingly pushes the boss to lead the conversation by forcing the boss to frame a go-forward position on the CISO's terms.

That's a lot of screwups in the first few minutes of an important conversation.

This is an age-old problem. As William H. Whyte said in *Fortune* magazine in 1950, "The great enemy of communication, we find, is the illusion of it."

Conversations like this often fail—or are off-kilter—due to skipping the negotiation phase of what the conversation is about or not coming to an (implicit) agreement about how to go about the conversation. Even "perfect" content can fail horribly without the proper setup.

So what does "good" look like?

Understanding and Leveraging the Implicit Social Contract

Taking the same scenario as above, let's change it a bit.

CISO: "I'm trying to learn more about the business so that I can better understand how my role as a CISO helps and gets in the way of the business, from other leaders' perspectives. Do you have time to answer a few questions from which I hope to get insights?"

Note: This isn't being crafty or manipulative. For example, the CFO doesn't typically go to the CEO to solve what are fundamentally finance problems.

Let's assume they say yes, either now or for another time. The conversation continues:

CISO: "Can you share your thoughts on what elements of operations are most central to the company?"

Don't fill in any blanks you encounter in your understanding with either assumptions or ego. If a piece is missing for you, or ambiguous, ask a follow-up:

CISO: "How do these areas support or affect the business?"

Boss: <the boss explains>.

CISO: "I see some ways, some more likely, that those processes could fail during various disasters. Who would be good to learn more from to make sure my understanding is correct?"

Then, following up:

CISO: "How should I initiate with them?" or "Could you provide an introduction?"

That's one example, but many approaches can work.

Breakdown of Why This Approach Works

No one can help you reach a conversational goal if you don't first propose one. Sharing your intent invites others to help shape the direction and purpose of the conversation, making it more natural, collaborative, and often more productive.

The most effective approach is to state your intent or goal before sharing information, starting a discussion, or asking for input, guidance, or decisions.

This is the beginning of "negotiating the conversation," which includes establishing why you're here and what each person hopes to gain from the exchange.

An implicit goal is to reduce the dissatisfaction and cognitive load of your conversation partner, and to yourself be open to where they want to take the conversation. Don't make them work to understand what you're trying to accomplish. Don't impose your goal on them and then make them lead. Guiding the conversation well and steadily is your job—even, and especially, with people more senior than you.

As Charles Duhigg writes in *Supercommunicators*, conversations occur across three zones: practical, emotional, and social. Most leaders are used to practical conversations, but few are adept at navigating the emotional question "How do we feel?" or the social question "Who are we?"

So how do we handle this?

Again, this leads back to breaking the conversation into two parts: intent and content.

The quiet negotiation of every conversation is what the conversation is going to be about.

So who do we have conversations with?

5.6 Map Stakeholders Strategically, Engage Intentionally

Stakeholder mapping is a tool for identifying and planning how to engage individuals, groups, and organizations that have a vested interest in an initiative's outcomes, a function's goals, or broader business objectives. This process begins by systematically identifying those affected by an initiative's success or failure, then considering their professional, career, and organizational roles and goals.

Understanding stakeholders requires delving into their inherent and situational concerns, and how your initiatives intersect with their interests and influence. Early stakeholder mapping envisions plotting stakeholders on a two-by-two grid (influence, interest), then designing tailored engagement approaches that include audience, key messages, cadence, and channels.

In a prime example of selecting well from a wide range of models, this is not the time to use standard models from project management or governance. The RACI Matrix (responsible, accountable, consulted, informed) and its cousins will fail to broadly help you in your CISO role, because as an executive you're not a "project" in the sense these tools are suited for.

A more relevant starting point for stakeholder taxonomy involves addressing the specific roles stakeholders play in relation to cybersecurity (see note box in chapter 2 for an example).

To navigate the dynamics of these stakeholder groups effectively, understand how each is affected by your programs, initiatives, goals, policies, or decisions, and find ways to make your work valuable or at least acceptable to them.

Once you've mapped these dynamics, it's crucial to understand the organizational decision-making models in play. Knowing whether you operate in a consensus-driven environment or follow an innovation pipeline process will help you anticipate potential roles, challenges, objections, and the ways stakeholders influence initiative approval.

5.7 Being Aligned Is More Than Talk; It's Practice

For a more structured conversational design, several methods are available, some of which you may already use for other purposes.

These real-time, hands-on, and collaborative approaches can foster a much deeper and more nuanced mutual understanding, but they require time and effort to implement effectively. Success depends on a deeper-than-usual understanding of the business and its priorities. To achieve this, CISOs can work with stakeholders across departments to gather insights on risks and impacts, engaging cross-functional teams to bring diverse perspectives to the activity and foster buy-in for the process.

In general, these are part of combining insights from broader business objectives with ongoing cybersecurity initiatives to help with buy-in and right-sizing resources from leadership.

The following common cybersecurity activities often overly focus on technical details, which can be repurposed to this end:

1. Incident Postmortems: To avoid being perceived as reactive, frame postmortems as opportunities for learning and improvement. Framing postmortems this way helps align leadership around the value of future-focused strategies.

2. Premortem Assessments: By envisioning likely-case scenarios and walking through how the group would respond, leaders can identify areas where they assumed someone else was responsible, identify gaps in their preparedness, and work to improve integration across departments.

3. Scenario-Based Threat Modeling: Engage leaders in scenario identification to make risks more concrete. Tailoring scenarios to reflect realistic threats and business-specific concerns makes the work of cybersecurity more relevant. The scenarios can then be integrated into ongoing design, planning, and risk communication.

4. Facilitated Risk Assessment: Risk assessments can be collaborative exercises through well-structured discussions between

cybersecurity teams and stakeholders. Business impact assessments are an example of this, unlike traditional technical assessments that are merely passed to the business. When stakeholders are actively involved, these assessments often yield valuable, eye-opening insights for everyone.

Think about the CNO from chapter 2. Initially, she was ready to tell her staff to do the opposite of "what HIPAA required" because it was contrary to what she cared about. I stopped talking about cybersecurity and instead listened and then presented a plausible scenario grounded in her deep concerns, and it changed everything. Critically, I stopped telling her how to do her job.

The four methods above are building opportunities for you to do the same. Each of these methods succeeds when you use them to connect what you do to what your stakeholders care about, rather than technical implementation.

Seeking meaning and living with purpose is a human universal— don't fight it. Use these methods to work with it.

5.8 You Can't Succeed if Others Define Your Success Differently

Most CISOs walk into a new CISO position thinking they know what the job is.

That's reasonable, right?

You know what your job is. After all, there's a title and a job description, and, of course, your past experience. And your boss hasn't told you you're doing the wrong work. Plus those handy cybersecurity frameworks that I already trashed—especially those frameworks I already trashed—just follow those.

And everyone in the organization knows what a CISO does. Right?

Not so much—on any of those counts. And most CISOs sense this gap in understanding and consensus on the role, even if indirectly.

So how do you address and leverage this gap?

Stakeholder-Centric Role Calibration Gives You Clarity, Permission, and Runway

You can't live up to dozens of nonoverlapping sets of expectations, some of which may have little resemblance to one another or what you expect of yourself, and expect to succeed.

Understanding stakeholder expectations and negotiating to be cohesive and reasonable is one of the expectations of an executive.

In this regard, the CISO role is where the CIO role was thirty years ago—relatively new, undefined, and misunderstood by everyone.

That's a risk and an opportunity.

To address this risk and capitalize on this opportunity, this section builds on earlier concepts, turning them into an approach, with implementation detail. We do this through a three-phase role-calibration process that consists of (1) a Listening Tour to learn about your stakeholders, (2) gathering and negotiating stakeholder expectations, and (3) seeking feedback for calibration and performance.

The approach is designed to be practical and, as an additional benefit, makes significant inroads toward trust-building and relationship-building. You can use these straightforward steps to build cooperative relationships, enhance leadership skills, and ultimately elevate the CISO's performance and influence within the organization.

To get the greatest benefits, success rests on several foundational mindsets:

1. Treat these conversations as learning opportunities, akin to a compressed MBA.
2. Follow your curiosity to discover the human side of leaders, to understand motivations.
3. Be open to challenge and feedback, which is the path that leads to growth.

This is done in three distinct phases, with nine steps, which should be done in order:

LISTENING TOUR	ROLE NEGOTIATION	QUALITATIVE STRUCTURED FEEDBACK
1. Professional Insight Interviews 2. Interpersonal Reflection 3. Identification of Mutual Interests	4. Need-Clarification Interviews 5. Broad-Ground Analysis 6. Negotiation of Role Expectations and Boundaries	7. Structured Feedback Interviews 8. Strength Elevation Goal Setting (Force Field Analysis) 9. Communication of Intended Changes

The three phases and nine steps of the stakeholder-centric role calibration.

The sequence of steps is intentional—to provide the most likely path to successful outcomes and to avoid confusion on the part of participants. Specifically, this is done by avoiding multipurpose conversations where intent is muddled, and also by ensuring multi-stakeholder inputs to subsequent steps.

The order does matter. To illustrate, what's the point of asking for feedback on performance (#7) if there isn't agreement about what is expected (#6)?

We'll start with the first step, which is to orient yourself to the needs and priorities of others within the organization.

Role Calibration / Phase 1: Listening Tour

A Listening Tour is a strategy-enabling and insight-gathering approach designed to foster your understanding of the organization's leaders, their view of their role in the organization, and the organization's leaders as a system that delivers on organizational goals. This means understanding what matters to them and how cybersecurity can be used to support their goals. It embodies a philosophy of actively engaging with stakeholders to comprehensively understand individual leaders' goals, challenges, motivations, expectations, and perspectives.

The Listening Tour process is crucial for any leader, but especially a CISO, who depends so heavily on others to fulfill their purpose and responsibilities.

The Listening Tour is also a foundational step in aligning cyber-security and the CISO role with the organization's culture, values, and priorities. By understanding each leader individually, you can apply systems thinking to see how these parts interact within the whole—a skill CISOs often excel at in other domains, provided they have sufficient insights into the system's pieces.

The following sample questions can guide these conversations:

1. What do you see as your role in the company?
2. What are the critical things that you do that contribute the most to the company?
3. What are your challenges?
4. What's a pain for you?
5. What do you wish could be solved once and for all?

6. What wisdom would you offer to someone starting out in your field?

Phases of a Listening Tour:

1. Professional Insight Interviews: Meet individually with each leader, using a prepared set of questions to guide meaningful conversations.
2. Interpersonal Reflection: Reflect on the insights gained immediately after each conversation, then review and analyze the collective input from all discussions.
3. Identification of Mutual Interest: Identify patterns and shared interests among leaders. Consider how these insights shape your understanding of the organization and your role within it.

Note: Avoid transactional and awkward "style" and "preference" conversations. Learn about style preferences indirectly, through this process, by noting stories of past conversations and interactions and how those conversations are characterized.

What to Expect

This process can take weeks to complete, sometimes a month, depending on schedule and availability. This step can be done at any time and should be done with each significant job change.

Most people who undertake this comprehensively gain confidence in some of their working premises about the business while being entirely surprised by what they hadn't understood—ranging from unknown key value levers to an initiative that nearly everyone wants to abandon, but haven't figured out how.

> **Note:** Sometimes, you'll uncover projects that everyone wants to stop, but nobody knows how to stop them. When this happens, you're in a unique position, as the CISO, to derail this type of initiative, which is an example of an early win or currying favor. But this is a dangerous tool—be sure you checked everywhere for political land mines before attempting this.

Be open-minded and try not to think about cybersecurity at this stage. That will come later.

Role Calibration / Phase 2: Role Negotiation

The second leg involves clarifying how others perceive the role of the CISO and building a consensus on what the role entails. Without this clarity, any feedback you receive will be muddled and unhelpful.

At its core, this is about gathering insights on how stakeholders view the CISO's role. It's not about dictating what you do, or justifying it, but rather learning how others perceive your responsibilities and where they believe your efforts should be focused. This understanding is foundational for establishing effective and cooperative relationships.

If you and the people around you don't agree on the role, you have little to no foundation for collaboration, success, or even professional feedback. If everyone expects something different from you, and it's not addressed, conversations will tend to be awkward, off-kilter, or jarring—and, most of all, frustrating for everyone involved.

- How do you view the CISO role?
- What do you need the role to do?
- What do you need the role to stay clear of?
- When should the role be in the room? When shouldn't it be? Why?

Be prepared to hear a wide range of views of the CISO:

- Ensure we're compliant.
- Make sure I'm not making mistakes.
- Protect us, but don't get in my way.
- Offer advice as part of my decision-making process.
- Provide insights to help me understand what I'm facing.
- Keep my team on track.
- Get me funding to address my issues.
- Give me a communication path to leadership or the board.
- Do your job so I can do mine.

There are many legitimate views, and many will have nothing to do with how you view the work your team does.

A Risky Line of Questioning: You've Experienced the CISO Before, So What About That?

You can learn from stakeholders based on what they liked and didn't about the way the role was handled previously. This can be done with carefully crafted questions that achieve clarity through contrastive role clarification but in safer framings:

- What expectations of this role have changed over time? (change framing)

- How did the position affect you before? (oriented-to-them framing)
- What do you want this position to focus on more? Less? (future framing)

Craft questions that work for you and make them yours, but start with the assumption that most experienced executives will prefer not to bad-mouth former colleagues. Some tout dicey questions as "direct" or "candid," but others may consider you rude for asking them to bad-mouth colleagues, and they may question your tact. In the end they rarely get you more useful information than the better-framed questions.

Personalized questions are generally the diciest and are best avoided:

- What did the last person do well? Do poorly?
- What worked in their approach? What didn't?

You use these at your own risk, and before you do, first carefully judge whether the person you're sitting across from wants this and has invited this—and, importantly, whether it's part of the organizational culture.

The value of these questions is not just in the answers; it's also a signal to the recipient that you are something different. Something more attuned and more adept.

Don't Push Back; Just Collect Data and Save Negotiations for Later

First, listen to everyone without arguing. You're not here to tell them what you do, or what you should be doing. You're not here to make decisions, or change hearts and minds. This is an opportunity to learn and for developing business acumen and productive relationships. And to reiterate an earlier point, this is not a RACI (responsible, accountable, consulted, informed) exercise.

This mindset is challenging but crucial. Separate the initial feedback meeting from a follow-up discussion to allow you to analyze all input and give yourself time to process. After the first round of discussions, summarize what you heard and present it in a second one-on-one meeting with each person. For example, if the CEO expects you to stay silent unless it's radically urgent, explain why you can't do that as a CISO and provide your reasoning, while tying it into what other leaders are expecting of you. In Chapter 6, I'll share an example of involving my COO in a similar way to address a tough situation.

Report Your Analysis and Build Consensus

After receiving input from everyone, report back. This shows that their input had an impact. The following are the core elements of doing this effectively:

1. Acknowledge. For example: "Thank you for the conversations about the role of the CISO and for your insights."
2. Address themes. For example: "I heard what you said about _____, and it was something that I heard several times, so I will be focusing on that. I also heard what you said about _____, and I will be looking for places where I can apply that as situations come up."
3. Share impact. For example: "I see why this is important, and I've already started to think differently about my role. Just yesterday, I _____, which was new for me. I expect as I think about it more, it will affect _____ and _____. Again, thank you."

This is illustrative, and it's important to use your own words and style. But cover these three aspects in your follow-up to ensure your input providers know their input is valued and acted upon.

Ask for additional input to further calibrate. This shows ongoing collaboration, listening, and thoughtful leadership, aligning everyone's expectations while making it clear you won't follow every suggestion but will consider all feedback to find the best path forward.

The underlying process is commonly known as shuttle diplomacy.

Once you have clarified and gotten consensus on the role, and only then, you have earned the right to ask for feedback.

Role Calibration / Phase 3: Qualitative Structured Feedback

The third phase focuses on gathering feedback on your performance, but only after establishing a clear understanding of your role. Differentiating between feedback on the role itself and on your execution of it is essential to ensure that feedback is relevant and actionable.

This phase centers on how others view your performance in the role as defined by the prior phase—not your personality, or how well you've generally "performed," but specifically against the role that you designed in phase 2.

Qualitative Structured Feedback

Recognizing the complexity and evolving nature of cybersecurity leadership roles, I developed the Qualitative Structured Feedback (QSF) process to help clarify key focus areas for developing executive skills and behaviors. This process uncovers the critical insights needed to effectively evaluate and enhance your performance.

By focusing on others' perspectives and fostering open communication, you'll gain a deeper understanding of your strengths and areas for improvement.

I recommend writing out the questions ahead of time and using the same questions with everyone. Consistency in the questions lends itself to more quickly and clearly identifying patterns you can then reflect on.

As an executive, you're expected to allow for a natural conversation, listening to meta-conversational signals and moving the conversation along as appropriate, which still gently guides the conversation where you need it to go.

Phases of QSF:

1. Interviews for Feedback Collection: Gather insights from close colleagues—peers, leaders, and clients who know you well and feel comfortable offering candid feedback. Ask specific questions to identify areas for growth and clarify any ambiguities. A diverse group is key for richer insights.

2. Reflection and Action: Independently or with a coach, analyze the feedback, categorizing it as "expected," "surprising," "unclear," "questions," or "actions." This reflection helps pinpoint areas for improvement and establishes a clear path for enhancing your leadership skills.

3. Communication of Intended Changes: Share your intended changes with those who provided feedback, expressing gratitude, summarizing insights, outlining planned actions, and inviting further input. Be clear about which feedback will be acted upon now and which may be considered for the future.

Again, as with all conversations, be sure to lower cognitive load for the people you are interviewing. Set up the conversation well. Talk about your intent, and let them know what you will do with the information.

Challenges of Access and Proximity Dynamics

Most modern executives have faced challenges of access and proximity dynamics of some sort during their careers: navigating through executive assistants appropriately guarding their leaders' time, working with remote international VPs, or wrestling with other obstacles to securing formal meetings or lucking into informal hallway conversations. These challenges have been with us to an increasing degree and with increasing prevalence from the first time a company had far-flung offices, likely from the time of Mesopotamia, four thousand years ago.

There will always be new wrinkles to this—whether it's hyper-exclusive C-suites (1980s), new ways of working during the rise of knowledge work (1990s), confusion over the etiquette of digital communication and collaboration tools (2000s), the growth of off-shoring (2000s), the challenge of looking too eager talking to the CEO in open office environments (2010s), or the complexities of near-universal remote work (2020s).

In each era, the challenge remains the same: connecting with the right people at the right time in ways that respect their roles and priorities using the tools and etiquettes of the period. Access isn't just about getting a meeting or having a face-to-face; it's about building the relevance of your work to what matters to them—executives and employees alike. Proximity, similarly, goes beyond physical closeness—it's about understanding their goals and connecting your leadership and efforts to their outcomes.

Executives who master this turn hallway conversations, digital tools, and virtual meetings into ongoing opportunities to collaborate, lead, and add value. In these instances, the channel and method still matter, but adaptability matters more.

Variations for Your Direct Reports

Among feedback providers, direct reports can be the most nuanced and challenging group to get candid input from. Without a steady history of showing openness—and, occasionally, in public or semipublic ways, taking responsibility for mistakes others point out—you'll likely struggle to get that door open once or twice a year. You haven't yet proved yourself to be a trustworthy recipient of feedback.

If this rings true, consider anonymized processes until you can build up your trustworthiness. While anonymized feedback might lack the nuance and clarity that comes from face-to-face conversations—and you'll lose the ability to follow up with individual providers—it can still generate workable input.

Assuming you have open lines of communication with your direct reports, seek their feedback directly. Start by clearly explaining your intent, how it aligns with your broader goals, and what you aim to achieve. Direct reports are often hesitant to provide honest feedback without understanding your specific underlying motives.

Elements to include in your setup:

1. State Your Intention: Let them know why you're asking for feedback. For example: "I want to make sure I'm not missing anything. You see how I operate, and your insights are valuable for helping me improve as a leader and better support the team."

2. Reassure Them: Address the power dynamic by saying, "I understand this might feel uncomfortable since I'm your boss. Your feedback won't affect how I view your performance."

3. Give Them an Option to Decline: Offer them the choice to opt out without pressure: "If you're not comfortable providing feedback, that's completely fine. I appreciate you hearing me out."

4. Acknowledge Your Own Vulnerabilities: Start by admitting a perceived flaw to set an open tone: "I think I may talk too long in meetings. What's your perspective on that?"

5. Ask Specific Questions: Begin with specific issues, then move to more general questions like, "What do I do that frustrates you?"
6. Set Realistic Expectations: Be up front that you may not be able to act on all feedback, but that you'll carefully consider all of it.

By structuring the conversation this way, you create a safe space for honest feedback, build trust, and ensure that the feedback process is constructive and aligned with your growth objectives.

Isn't That Servant Leadership?

This isn't servant leadership, which asks, "What areas are blocking you from getting your work done?" That approach is operational. Here the focus is on your behavior as a leader, not specific actions to support your team. Helping your team operationally is one thing; becoming a better leader is another. This process is about enhancing your leadership skills, not just removing obstacles.

Can't I Do the Three Steps All at Once? As a Group?

While it might seem efficient to gather everyone together, this approach is generally not advisable. By approaching this one-on-one, you're essentially engaging in "shuttle diplomacy," negotiating between different parties with different wants and needs, and using these differences to land in a beneficial place for all, including and especially yourself, while refraining from manipulation.

In rare cases where there's significant disagreement among key stakeholders and constant conflict, it might be necessary to bring everyone together to hash out differences, as I will explain I did with my chief legal officer (CLO) and COO in chapter 6. But you want to start with one-on-one.

A Better Definition of Early Wins

Everyone agrees it makes sense to score early wins starting as a new CISO.

For whom?

Many CISOs will say that an early win is doing an assessment of their department's performance and security maturity and developing a plan from that assessment.

That's like buying a house and telling your partner that your early win is assessing the room that will become your retreat room ("she shed" or "man cave") and building a list. How much of an early win does that sound like to your partner? Not much, I'm guessing.

I'll say your boss doesn't care to any great extent unless they already know there are problems and you were specifically hired to address them. The CEO likely cares even less.

What do they care about?

Your peers and your leaders mostly don't care about that stuff. What they care about is you removing obstacles and risks to their success.

I recommend to my clients that, during their Listening Tour while joining an organization, they ask about thorns others are experiencing. Michael D. Watkins, author of *The First 90 Days*, talks about this.

Ask them about those thorns, and then do your best to remove them.

In the eyes of your stakeholders, that's a much bigger win than yet another report that drives more work for their teams.

Give before you take. When you've made their lives easier, stakeholders are more amenable to what you have to say.

That's an early win with long-term impact.

Questions

1. What impact does the trust equation have on your approach to others in the workplace?
2. What adjustments can you make in your everyday conversations with non-cybersecurity executives to better understand their goals and challenges?
3. What steps could you take in your effort to "align cybersecurity to the business" to make it more concrete?

CHAPTER 6:

Strategy Creates Its Own Self–Calibration and Momentum

"Build our business continuity capabilities from scratch. Do it in one year."

One year before the HIPAA Security Rule compliance date, I was tasked with achieving 100 percent compliance with the disaster recovery and business continuity requirements of HIPAA, along with the HIPAA Security Rule standards and specifications, in one year. Starting with whatever the organization already had.

The CLO saw it as my sole responsibility and said so repeatedly at the executive-level HIPAA Security Rule steering committee. This led to significant strife on the oversight committee. My CIO and I both felt caught and somewhat disempowered for a month.

I mentioned this in one of the standing meetings I had with the corporate strategy director. She advised, point-blank, to schedule a meeting with the COO, and said she'd personally give the COO a heads-up.

While the COO was a standing member of the committee, it had been a change-laden quarter, and she hadn't attended in two months. From the advice I realized that my job wasn't to fight the CLO as the most-senior member of the committee regularly present; my job was to bring reasonable options to the table. I realized I couldn't accept a responsibility that would require the help of the COO without consulting the COO.

So, on the basis of groundwork and trust I had built with the COO, I scheduled a one-on-one meeting with her. I walked her through the situation, expressing that I viewed my authority, reach, and knowledge of operations as insufficient. She came to her own determination that business continuity was her responsibility, even while HIPAA Security Rule compliance was mine. We largely split disaster recovery planning (DRP), with her as the sponsor and me as the implementation coordinator.

Importantly, she knew the tenacity of the CLO. So she came to the next committee meeting and laid out who she thought was responsible for what, and she left right after, a mere five minutes into the meeting. The topic of core accountability and responsibility for business continuity planning (BCP) and DR never came up again.

The company also later hired a BCP manager after the COO realized that she didn't have time to commit to driving it on her own.

The lesson here is that CISOs cannot resolve organizational challenges alone, especially when navigating the complex dynamics between powerful stakeholders. It's about leveraging standing meetings and processes, leaning into the trust you've built, and getting key executive support to facilitate solutions, rather than getting entangled in intractable conflicts, acting as a proxy for someone who's not in the room.

This approach can be applied to cybersecurity decisions of all sorts and sizes. In essence, the one key facet of the effectiveness of a CISO is their ability to foster clarity for accountabilities and shared responsibility across the executive team.

6.1 Strategy Is Not What You Think It Is

I recently discussed strategy with a seasoned cybersecurity executive, aiming to clarify what strategy means for cybersecurity and technical executives. Here's a condensed version of our conversation, edited for clarity and brevity.

Clear the decks. You're about to get something new.

Richard: What is strategy?

Chris: Strategy is one of the most confused concepts and misused terms in business. Contrary to common practice, developing a strategy is not selecting a list of priorities, developing a plan or roadmap, setting goals, adopting a framework, or preparing for the budget process.

Richard: Okay, but what *is* strategy?

Chris: Strategy is about choices that form a unique approach to challenges. Roger Martin describes it as an "integrated set of choices" for positioning to be the best in your space, which includes defining your space. Richard Rumelt calls it a "coherent mix of policy and action" to overcome high-stakes challenges. Both definitions converge on purpose, framing addressable challenges, and developing principles.

Richard: What does *addressable* mean?

Chris: An addressable challenge arises from the specifics of your situation, is counter to your purpose, and is within your control. Complaints like "we're not able to mandate controls" and "the business doesn't care" aren't directly addressable; they're grievances, not challenges. You might as well stomp your feet and cry. Challenges can't simply be something that you want.

Richard: So what does addressable look like? Can you give an example?

Chris: Let's consider a common cybersecurity complaint: "The business doesn't follow policy for business-managed IT." Often, the response from cybersecurity or IT executives is to demand compliance. But that approach rarely succeeds and almost never drives the kind of engagement or behavioral change needed for sustainable improvement.

This complaint usually stems from one of two root issues, though it may vary by organization. For the sake of example, let's assume it's either because the business believes its practices are "good enough," or it's avoiding the perceived "cost" or "friction" of centralized IT.

Understanding which of these is driving the behavior will change how you approach your principles.

In the "lack-of-understanding" scenario, you might establish a principle such as "provide for accurate self-evaluation of implementation." The goal here isn't to train them to "do it right," which doesn't address motivation. Instead, the principle subtly encourages an honest assessment of where they stand. Culture is crucial in this scenario. Whether the department values competitiveness, quality, or innovation, shape your principle around the group's intrinsic motivations. This might mean appealing to their pride in excellence or competitiveness, rather than assuming they simply lack best-practices knowledge.

The issue here isn't a lack of knowledge; it's an unawareness of their own gaps. Framing the challenge as "they're unaware of their gaps" is far more solvable than "they need training in best practices."

In the "cost-avoidance" scenario, your principle might look like "surface hidden costs," with the objective of adding visible "costs" to business-managed IT. The aim isn't to prevent the business from handling its own IT, but rather to make it a less attractive option. For example, you could expose hidden risks by internally sharing the gaps in business-run IT or even linking these gaps to higher insurance premiums—there are numerous ways to address this.

In both cases, these are cyber-led (or potentially IT-led) actions, not demands on the business. Well-formed challenges avoid the mindset of "it's their fault" or "if only they would just."

Richard: I would have thought that the strategy was the mission. But you're saying that you've got your mission and purpose, so what's stopping you from fulfilling it, and bringing those together to form your strategy?

Chris: That's where it starts. Purpose and challenges form the crux of the problem you're trying to solve, and the solution is embodied in principles. You can think of principles as philosophy-level rules that apply to yourself with the goal of influencing others.

For example, in situations where the business generally wants to "do things right" but doesn't fully understand what that looks like, the aim is to guide them to recognize their own limitations—not to micromanage or "correct" them. In this case, you're helping the business see where it stands rather than dictating how they should get there.

This approach requires training IT staff to educate non-IT stakeholders constructively, showing them what needs to happen to achieve strong operations and cyber-security. Instead of creating a new cyber function, this principle naturally leads to a new engagement model for every IT-business and cyber-business interaction.

The bet—or gamble, since strategy always involves an element of risk—is that the business will either choose to leave IT and cyber matters to the experts or will commit to doing it well themselves.

Either way, the principle holds: Equip those responsible for security to know good practices from bad. The goal isn't to control every detail of IT or cyber, but to achieve an outcome that aligns more closely with your mission and purpose.

So the key question becomes: What do you actually have control over?

Richard: Mostly, when I hear about strategy in businesses, it's all about the market and external forces. But you're looking inward. I'd assume strategy should be based on protecting the environment, but you're not saying that at all?

Chris: No, it's not as broad as a business strategy.

But it's also true that, as a profession, we're not thinking strategically enough. We're not challenging ourselves to redefine our roles or frame our challenges broadly enough. Too often, "strategy" in cyber means trying to gain control—when, in fact, letting go of control to gain influence would yield far better results, perhaps even double-fold, and would be more sustainable in the long run.

Principles like this do more than just tell people what to do—they change how people think and approach their work.

Richard: Let's say you've had multiple breaches. That's a serious problem, and the business wants to stop it. Could you come up with a strategy to fix that?

Chris: Breaches are an outcome. So let's explore. With the starting point of having had several breaches, what led to those breaches?

Richard: It could be a lack of training. It could be a conflict between departments. IT thinks that it's handling it, but it's not. It could be that there's not a process in development to make sure the code is secure. It could be that the routers or firewalls aren't properly configured. It could be any number of things. The databases aren't up to date, so they're getting SQL injections. All these things could be happening simultaneously, if you don't have the proper things in place to make sure that these things aren't happening.

Chris: So let's step back one order of magnitude. What is the assumption or assumptions that led to that circumstance of people not patching, or people thinking they're okay on security, or the other situations you mentioned?

Richard: What I've seen is that people think it's not their job. They don't know what they're doing. They don't understand the importance. There are no corporate goals at all about security. There's the sense of "we need to be secure," but there's nothing else. They just don't do it.

Chris: The crux of the problem is the thinking or point of view that led to those circumstances that led to a breach—not the breach itself.

Richard: So the strategy can't just be about finding a technique or product to fix the problem. That's not strategy—that's just a technical solution. And even if it solves the immediate issue, you may not even know the problem.

Chris: You're 90 percent of the way there. Let's keep going.

Let's say you've developed a set of principles.

Let's say one is to "align to the business." That's a terrible principle because no one knows what it actually means. We hear that all the time in cyber. But that has no bones, no meaning.

Pressure-testing this principle reveals why it's flawed. No one is quite sure if they did it or not, even though it sounds nice. It's a platitude, not a principle.

Something more concrete could be: "Use the business language of the audience." That shifts us.

For example, when communicating with a VP of customer service, instead of summarizing technical vulnerability reports, you might frame them as reports on potential business process failures.

If it feels like it could show up on the bottom of a 1990s motivational poster, it's not a principle.

Richard: AWS has a principle that security is priority number one.

Chris: That's not a principle—it's a platitude, a mantra, a tagline, or maybe a market position. At best, it's an internally weak call to action. The range of interpretations is so broad that it fails to drive consistent decisions or behavior.

Worse, the phrase "security is priority number one" ignores business and existential realities—most people won't believe security should come before staying in business.

A more practical articulation of a related concept could be "We do not compromise security to make a sale."

The goal is to create principles that function like inviolable policies—no one should want to violate them, and there should be no need to.

Take "use business language"—it means speaking in the business language of the audience every time, not just when it's convenient or for a single report. The whole department has to do it, consistently, in all communications. That might mean focusing on operational impact for a VP of customer service or offering code-pattern explanations and sprint-related approaches for programmers.

THAT is something people understand.

What comes out of deep strategy discussions are not only the principles themselves but also how they'll be applied, which in turn helps refine them. So "use business language" might quickly evolve into two principles: "speak in the audience's language" and "anchor to how the business defines impact."

Great leadership teams invest considerable time getting these principles right, because once they realize the power in this approach, they understand it's worth the effort. They know they have to get it right.

Richard: If developers were coding in production, we might make a principle that there's a wall between them, and there are no exceptions, aside from emergencies.

Chris: Not quite; that's just a policy.

A strategy principle might be "code deployment processes do not put production at risk."

A principle shapes how people think, rather than

simply telling them what they can or can't do. It provides guidance for making decisions. In this way, the implications and impact of the strategy are broader and more significant.

Interestingly, in the case of an outage that a code push would fix, it nearly spells out a quick, pragmatic code review rather than a full code review, to get the site back up and running.

There's nuance and subtlety here.

Richard: So, at the top, we have our purpose or mission. Then we identify the challenges to achieving that purpose. To overcome those challenges, we define several principles. These principles touch and impact everything we do.

Chris: Yes. That's right.

Richard: I have an example that might fit, but I want you to weigh in, because I'm almost getting it.

Early in my career, we had this issue where users would call in with problems, and we'd have to rush around fixing them like chickens with our heads cut off. We didn't have a ticketing system, so we decided we needed to record incidents. We didn't care who recorded it—the user could log it, or we could log it—didn't matter, as long as it went into a ticket. How would I turn that into a principle?

Chris: You wouldn't—because it's not really a strategic concern. Strategy isn't about figuring out what you are or aren't doing and then justifying it. Strategy is about identifying core challenges and developing foundational ways of thinking to address those challenges.

In this case, the principle might be framed as "have a basis for improvement."

This construction of a principle gives us almost automatic, solid judgments and strongly suggests a direction. Once that principle is in place—wherever there's a problem—it becomes natural to use the principle. It becomes natural to document issues and conduct after-action reviews—actions that support improvement.

A principle guides the direction without dictating exact steps, allowing the situation to determine implementation while shifting behavior in the right direction.

In your example, creating a ticket simply becomes a natural consequence of applying the principle "have a basis for improvement."

Richard: I see—I'm working in the wrong direction. I'm going up and I should be going down.

Chris: Yes.

That said, it works as a stack. You work downward for development and upward for pressure-testing and refining until it all holds together.

One litmus test is that the principles should be self-evidently beneficial to those who know the environment. For example, let's say you have a principle like "have a process to improve." From an organizational change perspective, who's going to generally disagree with that? And if the challenges are well understood, this principle should get people excited because they can finally see a way to address issues that have been frustrating them.

Now consider how broadly the principle "have a basis for improvement" could apply and how it might change mindsets, behaviors, assumptions, system design, and priorities.

Richard: So if I understand, you're going from strategy down to tactics?

Chris: I tend to avoid the word *tactics*. Most of the time it means whatever someone doesn't think is strategic—or vice versa. That's not useful.

Beneath the principles, we have strategy-supporting initiatives and strategy execution.

Initiatives are the actions that support the principles, ensuring they're applied effectively. For example, to make "use business language" a reality, you'd need to develop business acumen within your team, possibly through targeted training. This is just one of several steps that might be necessary to bring that principle to life.

Once enabled in this way, people are able and much more likely to live by the principles.

Through this training, team members naturally begin to apply the tools and ideas available to them whenever they see how it benefits their work.

But remember, you selected specific training, not just general training, because that specific training was necessary to make the principle actionable. That's strategy implementation.

Eventually, it all comes full circle. They'll start implementing the principle in both big and small ways in their day-to-day work. In this conception of strategy, people don't stop doing their jobs; they just start doing them in slightly new, principle-driven ways.

In the end, your department is going to modify everything they do.

Richard: I have an example: World War II. The strategy was "Germany first." That was it. That was the strategy. And then Normandy and Overlord became the tactic. Or a lower-level strategy.

Chris: That's a solid example. The generals and admirals debated that fiercely.

Notably, it demonstrates what strategy often does: It forces leaders to make hard, sometimes uncomfortable decisions up front to achieve cohesion—that is, avoiding small decisions that later work against each other.

They faced tough questions, like "Do we, as the United States, wage a two-front war?" versus "How can we let Japan keep doing what it's doing?"

With a "Germany first" strategy, the admirals were largely sidelined to support a land war, playing a supporting role rather than leading. That's a tough choice because it's difficult for action-oriented leaders to accept a reduced role.

In the end, the decision was a definitive "No" to a two-front war. It shaped the entire approach, down to the last detail, and proved decisive. So, yes, you're right in identifying "Germany first" as a core strategy principle.

Richard: Another principle that supported the strategy was "surrender is unconditional: We do not accept anything less than unconditional surrender." It caused the war to last longer, and it got bloodier, but it affected the prosecution of the whole war effort.

Chris: Yes, that's right. And that's another tough choice—more people died earlier in pursuit of that principle.

Together, those two principles—"Germany first" and "unconditional surrender"—are what strategists call *coherence*. They work together to make sense and drive more effective decisions about what you do and how you do it.

Richard: I didn't mean to go off topic, but as you were talking, it clicked that this is a perfect example of a strategy that actually worked and flowed down, with very few exceptions.

I'm curious, going back to CISOs, is this a strategy that the CISO should get approval for?

Chris: A CISO might discuss strategy in general terms with a few key stakeholders to refine and sharpen the principles. It's not about approval; it's about calibration.

Take Apple Inc. They have an internal strategy that most consumers feel without ever needing to know what it is—an emphasis on user experience. I don't need to know Apple's strategy to decide whether I like the result and want to buy an Apple product.

Similarly, a CISO will be making significant but subtle internal changes that don't require large budgets, per se. The impact on stakeholders is real but largely intangible.

You don't need to explain how these activities will feel different because the change is in the experience stakeholders have, not in how the CISO explains it.

Richard: It's basically a philosophy. Like "Germany first."

Chris: Yes. It's a philosophy applied to a specific situation to tackle a specific set of challenges.

Richard: This is so different from what I've been taught. I was always told strategy was about budgeting, figuring out how to get from A to B, or what we're doing this year—your typical timelines and plans. That's what consultants always called strategy, but it's not what you're describing.

Chris: That's what implementation consultants call strategy. But if you talk to big-firm strategists—the ones known as strategists, not just consultants who sell services—they'll give you a description much closer to what I'm talking about. It won't be the road-mapping that often gets labeled "strategy" in an implementation context.

Richard: This is starting to make sense. Without strategy, your road maps don't hold up because you lack that overarching set of principles. You end up wandering all over the place based on the market or the latest emergency—you know how it is.

Chris: Yes. And without strategy, your budget is just a junk drawer. There are things in it, but that's about all you can say.

Now I have a question for you. If you'd been introduced to this concept of strategy ten or twenty years ago, what difference might it have made for you?

Richard: Night and day. I didn't have any of this back then.

Chris: This is where strategy gets interesting. What would have been different?

Richard: For example, with disaster recovery, I was often told it was my responsibility, so other departments would just wash their hands of it.

But if I'd taken the approach of saying, "I don't care about DR; it's for you," I could have shifted the conversation. I could have asked, "What do you need from DR? How can I help? How can I facilitate you?" And that did change the conversation in a few cases where I tried it—accounting got involved, and operations became real partners.

Chris: That's a powerful shift. Now imagine applying that principle universally, and get everyone on your team to do it as well.

Richard: That would have changed things. But the issue was that higher-level leadership didn't share that concern, so I would've had to escalate to the C-level.

Chris: Or maybe you wouldn't. I doubt accounting or operations asked permission—they didn't go to the CEO asking, "Can we protect our stuff?"

Richard: That's true.

Chris: Where else could you apply a principle?

Richard: I was leading in an environment where everything was always urgent and everything was always an emergency. We could have put a principle in place to say that's not the way we ought to operate, that we want to operate on more of a routine basis.

Chris: That's a wish or a goal. It's something you can't directly control if driven by the larger environment. To make it strategic, the principle would need to happen upstream or be designed to shape behavior.

Ask yourself: What mindset, behaviors, or approaches could you change? What could you address without needing others to change first?

An addressable challenge means it's something the person whose strategy it is—*you*—can act on.

Here's a business example of an un-addressable challenge: "There aren't enough buyers."

Richard: So, for example, the business unit developers didn't include security in the development process. In my position as director in IT, I would have had to come up with a possible principle that includes the development person and get her buy-in on it.

Chris: "Create a solution together" relies on her and therefore isn't a strategy.

Let me ask you this: If the problem were solved today, what would you have done five years ago, without forcing others to change? What could you have put in place to shape the environment so that the problem you're facing today was less of an issue?

Richard: It's clear to me now that others should come to you advocating for budget additions based on risks or needs you've communicated, rather than you having to request budget changes.

While there may be exceptions and some negotiations required, ideally, your effective communication should make the need for these additions obvious to decision-makers. If you set up everything you said, appropriately, you shouldn't have to go demand a budget. If this isn't happening, it suggests a shortcoming in your approach.

I wish I had known this long ago. It would have changed everything.

Special thanks to Richard for indulging me in this conversation for the benefit of my readers. Let's do a quick recap.

6.2 If It's Only in Your Head, It Isn't a Strategy

The humorist Garrison Keillor opened his radio show with a skewed portrait of an imagined place: "Welcome to Lake Wobegon, where the women are strong, the men are good looking, and all the children are above average." It's a funny line, but it also nails a deeper truth: Ideas that live only in your head, untested and unchallenged, can feel real and concrete while being utterly disconnected from the world around you. Beliefs and assumptions, no matter how tidy or compelling, stay exactly that—beliefs—if you don't put them into practice. Concepts that are untried and internal don't magically transform into strategies.

6.3 Strategy Is a Lens for Making Decisions, Not a Checklist

Strategy isn't a static set of steps to execute. Instead, it's a dynamic lens that refines how you view and respond to situations as they unfold. Unlike a checklist, strategy isn't designed to deliver certainty but rather to support ongoing learning and adaptation. It acts as a continuous guide, helping you evaluate decisions in the face of new insights and shifting conditions, which makes it a living framework rather than a finite task.

6.4 A Shifting Strategy Means You Have No Strategy

The dirty little secret of the strategy industry is that there's no process for strategy. The best you can do is have a strategy for strategy.

Strategy does not arise from a repeatable, linear, fixed process. Indeed, if this was your process, the result is not a strategy. Likewise, it is not a free-flowing, unstructured, loosely intentioned process. It's about an intentional approach that moves from informal ideas to formal principles through dedicated effort and refinement, and it's hard work.

From our discussion, we concluded that strategy is not any of the following:

- Aspirations
- Goals
- Plans
- Priorities
- Budgets
- Demands

The following are also not strategy:

- Vision
- Metrics
- Risk Management
- Business Cases
- ROI Calculations

None of these items create strategy. They are components of plans, goals, or supporting metrics. Strategy isn't future-oriented in the sense of achieving specific milestones. Instead, it's a steady framework that should rarely change—perhaps once a decade in established cyberse-curity teams within established organizations.

6.5 Strategy Is What You Influence, Not What You Demand

The strategic challenge for security isn't to convince others to give us more money or resources. Those are indirect complaints or direct wishes, more than an approach to strategy. Getting resources is a time-eternal challenge for nearly everyone, which doesn't point us toward what we must do.

Effective strategy often comes from letting go of a need for control and, instead, focusing on shaping perspectives, relationships, and environments. By recognizing that influence is more powerful than demanding control, strategy shifts from a struggle for power to a steady march toward fulfilling purpose.

Exceptional leaders focus on what they can influence, not what they can demand—they take decisive action and never blame others.

6.6 Strategy Is Grounded in Reality, Not in Ideals

Strategy emerges from seeking to understand, not sidestep, the constraints and opposing forces that you face. It's convenient but lazy to frame strategy as the pursuit of ideal states (e.g., "great security," "full compliance," "the budget we 'need'").

Effective strategy draws its power from the realities and challenges of your current environment, whether those are resource limitations, cultural resistance, or something else. By embracing and using these constraints, you're less likely to base decisions on assumptions, and less likely to develop weak spots or swing from excitement to disappointment with every partial success or failure in attempting to achieve the ideal.

6.7 Strategy Emerges; It's Discovered, Not Planned

As illustrated in the discussion, strategy emerges at the intersection of opposing forces—between where we want to go and that which is in opposition to the progress to get there. Challenges arise in the unique ways these forces play out within the specifics of your current circumstances, culture, and goals. Strategy, then, is how we act on the insights these challenges reveal.

Principles, formed from these insights, are how we embody, communicate, and bring our strategy to life.

Think of strategy less as something you plan and more as something you uncover or discover.

Discovery speaks to both the process and the feeling of strategy coming to light.

Committing to a strategy can feel exciting, especially when you're likely already operating under loosely formed principles that align with your strategy discoveries. The easier half of strategy is solidifying these ideas that have been percolating.

However, committing to a strategy can also be uncomfortable, often requiring new and unfamiliar activities while letting go of others that, while interesting, may no longer serve the direction you've chosen. This shift may call for developing skills outside your comfort zone or accepting team attrition as some members find they do not fit the new direction.

Changes might not happen immediately, but as counter-strategic elements phase out, the coherence and clarity required for strategy increase.

6.8 Strategy Doesn't Care About Budgets and Approvals

It may sound tautological at this point, but it's necessary to say: Strategy can be executed immediately, continuously, and across every aspect of an endeavor or corporate function. Strategy starts today. It's pervasive, touching everything, without requiring a big budget or publicity. Beyond your team and a few key stakeholders for calibration and support, it doesn't even need to be discussed. Strategy may influence budget, but it doesn't require a big dollar spend.

When was the last time you encountered or developed something that does that?

Questions

1. How have assumptions about your role limited your strategic autonomy and effectiveness, and what steps will you take to correct them?
2. What complaints in your current situation can you turn into addressable challenges, and how can you ensure your actions remain within your control, without relying on others?
3. How do you distinguish feedback on your role from feedback on your performance, and how will you use this distinction to improve?
4. How does the idea of strategy emerging from opposing forces reshape your current approach?
5. In what ways might a fixation on ideal outcomes or "perfect security" be creating weak spots in your strategy? How can you reframe these aspirations to focus on achievable progress?
6. Where in your current role can you reduce reliance on formal authority to gain traction with your initiatives, and what new skills would support this?
7. How does viewing strategy as "discovered, not planned" open you to new insights from your team, and how might this change your approach to leadership?

PART III:

Development

Our goal for part 3 is to apply the concepts, insights, and new habits you have discovered—and will continue to curate for yourself—in a way that sticks.

No doomed-to-fail New Year's resolutions allowed. No feats of willpower. No distant, lofty goals.

You may have noticed that I gave you stories in previous chapters. Now the story creation gets turned over to you. With the tools and ideas in this book, you've already started writing your future. In this next part, you'll continue that work—discovering your purpose, steadily building toward it, and calibrating yourself to reach your goals and aspirations.

Build deliberately, steadily, and reliably into the future. Develop ways to layer in concepts, act on key insights, and build sustainable habits.

So far, this book has introduced important ideas. Part 3 is where those ideas come to life.

Three Essential Challenges You Face Now

You have three questions before you:

1. How do you keep your goals top of mind to find the motivation to continue?
2. How do you coach yourself using the tools of an executive coach?
3. How do you hold yourself accountable in a way that works?

There's no shortage of techniques to choose from: time blocking, journaling, checklists, habit triggers, accountability partners, meditation, reminders. The next three chapters will help you explore what works for you and discover your unique approach.

CHAPTER 7:

Purpose Drives Subtle Yet Powerful Moves That Guide You and That Others Can Embrace

This chapter is about finding the motivation and energy to break free from your "default" mode and move toward your purpose with greater speed and clarity.

The "default" is the passive trajectory you follow if you don't take time to examine what holds you in place. Moving away from this trajectory means actively choosing change and following through with sustained steps—steps that may feel uncomfortable but are necessary for meaningful progress.

To find deeper wells of motivation, start by assessing yourself: your purpose, values, mindset, assumptions, passions, and pet peeves. These aspects aren't just traits. They're signposts—guiding you toward what you want to achieve, what you value, and the self-imposed barriers holding you back.

Your purpose is foundational. While purpose and planning must come together, purpose comes first.

If you're not ready to face tough questions, build a foundation for executive development, or challenge how you see your value, I invite you to put this book down—but pick a specific date to come back to it. This isn't an indictment; it's simply practical. Do what's the best use of your time right now, given your goals.

If you picked up this book out of curiosity, to seek new ideas for their own sake, or to confirm what you already believe, then part 3 may be a solid review for you. The same goes if you already have dozens of tools for habits, productivity, and communication—and you're adept at using them and satisfied with your results.

If you haven't explored these things deeply, you're about to be introduced to many new ideas.

Why Was I Just So Direct?

I'm not here to push you—I'm here to free you from unnecessary constraints. Reading part 3 without taking action is a sure way to add another item to your mental "should do" list, and for most CISOs, that list is already overwhelming.

Consider someone who says they want to eat healthier but has doughnuts every morning and ice cream every night. You can't do both with integrity.

Life is about making choices and owning them. Excellence comes to those who take responsibility for their path—so choose one, commit to it fully, and make the most of it.

Note: If you're ready and want specific tools, try the workbook at **https://newcyberexecutive.com/cisoimpact-resources**.

If you want personalized but general help, consider a professionally experienced and qualified executive coach by checking what your company already offers. (Download the executive coach selection guide for CISOs at **https://newcyberexecutive.com/cisoimpact-support**.)

If you're ready and want a coach who understands your unique journey, that of a cybersecurity executive, who is also a qualified coach, contact me at **https://newcyberexecutive.com/cisoimpact-call**. We'll figure out what will work best for you.

Congratulations for continuing your journey, and thank you for letting me undertake this journey with you through this book.

How My Clients Grow

My clients' transformations lead to a sense of calm and equanimity toward what's beyond their control and an unshakable confidence in decision-making. They stop fighting battles that aren't theirs and shift focus to areas where they can create change. Old complaints don't resurface as frustrations—they transmute and dissipate.

It's incredible to watch, and I'm grateful to be part of and witness these journeys.

As cybersecurity executives, my clients want to protect their organizations, do what's right, and minimize the impact of any breaches. They face considerable technical, operational, and business environmental challenges.

Yet the greatest hurdles they face as leaders often involve their own thinking: conceptual limits they place on themselves, misconceptions about how others see them, and a lack of clarity about the role of a cyber executive.

When they address these gaps, they no longer blame others. In ways they never imagined, cybersecurity becomes something for the organization to solve, not just the CISO. This transformation happens through the methods expressed earlier in this book, but also through the wisdom that is gained through the process.

That's a very different place to be.

The following sections outline this work, which can help bring you closer to what my clients experience.

Understanding Self and Others for Transformative Change

The most transformative shift my clients make is developing a genuine understanding of others' perspectives, which in turn shapes their understanding of themselves.

Unexpectedly, understanding others allows them to show up more authentically, with greater authority and confidence, and this leads to a happier, more fulfilling professional life.

Paying Attention

An example:

Many of my research subjects and clients attend team or cross-functional meetings as required by their managers.

A surprising number of these CISOs report that, unless the topic is directly related to cybersecurity, they multitask—checking emails and handling operational issues instead of engaging fully.

If this is you, consider what you're signaling. By disengaging, you're broadcasting that you don't care about the business, only about cybersecurity. It implies you're not ready for a more influential seat at the table, because you're focused solely on your own domain.

To be seen as an executive leader, commit to being present. Contribute. Show genuine interest in topics outside your immediate area of expertise. Engage as a member of the CEO's team, or whichever team is in the room.

Imagine if a field goal kicker stayed in the locker room until needed rather than being on the sidelines. That kicker wouldn't last long on the team.

Don't be the CISO who "waits in the locker room."

Going Beyond "Because Cyber" and "for Cyber"

When I speak with CISOs about managing relationships with key stakeholders, I often find their approach is ad hoc. There's no consistent plan, cadence, or intent.

Regular meetings, if they happen, are often at the behest of a manager who expects periodic check-ins across the team. Sometimes it's a habit schedule inherited from a previous leader. The result is a lack of intentionality in these relationships, with the CISO often going through the motions.

Two major issues arise:

1. Inattention: When the CISO isn't fully engaged in the meeting, other participants notice and question why they're even there.

This erodes credibility and pushes people away, which is exactly the opposite of what most CISOs aim for.

2. Purpose: CISOs are invited to these meetings for reasons that go beyond cybersecurity. It's the CISO's job to understand what's expected of them—and it often isn't just to "wait" for a cyber topic to come up. The skills and insights CISOs bring to the table can apply far beyond cybersecurity.

Contributing as a Business Executive (When out of Your Depth)

One hallmark of a business executive is using your skills to contribute to others' goals. Here are some questions to guide your approach and contributions:

- How can my unique approach to problem-solving support the team here?
- Beyond cybersecurity, what do others say are my strengths as a leader?
- In other areas of my life (civic, family, etc.), what can I reliably be counted on for? How does that translate to the current moment?

I encourage you to think about these regularly for several months. You can also benefit from asking your circle of advisors for their answers to these questions.

In the meantime, if you're unsure how to contribute directly, you can still add value by making sure the right questions are being asked. Here are two frameworks to move from domain contributor into meta-contributor.

Six Thinking Hats

If you have no other way you can contribute directly to the conversation, be a meta-contributor. One model for this is Edward de Bono's Six Thinking Hats. These hats cover the following: gather comprehensive information (White Hat), explore benefits (Yellow Hat), assess risks (Black Hat), check emotions (Red Hat), encourage creative solutions (Green Hat), and manage the process effectively (Blue Hat). Develop your own personal set of questions for each hat and keep this in your back pocket. (Sample questions in appendix D.)

Beautiful Questions

In *The Book of Beautiful Questions: The Powerful Questions That Will Help You Decide, Create, Connect, and Lead*, Warren Berger suggests "good questions are propulsive, they energize us instead of slowing us down or paralyzing us with doubt." For self-development, he proposes five questions that leaders can ask themselves. You can begin by asking these questions, or by setting up others to ask these questions:

1. How can I see this with fresh eyes?
2. What might I be assuming?
3. Am I rushing to judgment?
4. What am I missing?
5. What matters most?

> **Note:** Use intersections to generate innovation. Most innovation happens at the intersection of two fields. How do you use your cybersecurity concepts and experience to enhance

the work of others? How do you drive innovation and speed from your experience?

CISOs are good at wearing de Bono's black hat by default, as risk professionals. How can this fine-honed sense for what can go wrong be applied constructively for topics outside of cyber?

7.1 Don't Be Your Own Obstacle

Throughout our lives, we unconsciously absorb messages that influence our decisions and actions. We may look back and struggle to understand why we made certain choices or behaved in specific ways—often without realizing these patterns are rooted in early beliefs.

Sometimes these beliefs become so ingrained that we stop questioning them, and when things go wrong, we may blame luck, internal flaws, or even others, rather than examining their true origin.

For example, someone who grew up hearing "Wealthy people are evil" might find themselves overspending as an adult without realizing that they're avoiding wealth so they don't become "evil." They recognize the habit of overspending but can't connect it to the early belief, leading to frustration without clarity.

What areas keep tripping you up, but for which you don't have an explanation?

For many CISOs, executives, and people in general, these unconscious beliefs can influence behaviors in many areas:

- Value & Time
- Communication & Influence
- Authority & Power
- Service & Sacrifice

How do early beliefs shape behavior today? Here are a few common examples:

- "People with power abuse it" may lead to self-disempowerment.
- "Idle hands are the devil's workshop" may result in constant busyness without reflection or planning.
- "You're good at watching out for others" may create a protector complex.
- "You're so quiet and well behaved" may hinder speaking up when it counts.

Mental Health

It's important to acknowledge that many adults, including leaders, carry these experiences, and that most adults, including executives, are experiencing or have experienced depression, anxiety, bipolar disorders, substance abuse, or trauma.

It's real. I raise this for your awareness, and to personally acknowledge you, if you experience any of these.

If this is relevant to you and you are interested in executive coaching, I encourage you to find an executive coach who has trained in these and is familiar with complementing or working with mental health professionals to be a part of your team. While coaching doesn't address the source or resolve the underlying situations, a coach can offer structures and techniques that complement therapeutic approaches.

Importantly, if any of these are of concern, and if you have not already, I encourage you to seek out a qualified mental health professional.

7.2 Awareness Doesn't Equal Understanding

Awareness of leadership tools—like communication strategies and decision-making frameworks—is just a starting point. Awareness is passive; it involves recognizing tools without actively applying them. In contrast, making these tools effective requires active engagement, where leaders skillfully adapt and apply them in real-world scenarios.

For leaders, this distinction is critical. Knowing about a tool versus using it effectively shapes your impact on decisions, influence, and outcomes. Many executives are aware of leadership techniques, but without the ability to effectively use them, these tools remain theoretical. Effective leaders don't just have a tool kit; they have the intuition and experience to select and use each tool well, for any given situation.

The difference between knowing and doing is the difference between potential and impact—between leadership in theory and leadership in action.

In the pages that follow, I introduce several tools that are useful on their own, but I have selected them for the mindset they embody. In this way, you are learning not only a tool but also a way of thinking. Even if you never use the tool, the mindset remains available to you.

Clear Communication Without Assumptions

Effective communication focuses on your own experiences without projecting assumptions onto others. This approach helps prevent misunderstandings and conflicts by respecting personal boundaries and refraining from unintended judgments. This style is especially valuable for managing the emotional dynamics in most conversations, keeping interactions open, respectful, and constructive.

To communicate effectively in sensitive situations, center your statements on your own observations and the impact on you. Avoid framing your message in a way that assumes others' intentions, feelings, or thoughts. This minimizes defensiveness and preserves the integrity of the conversation, fostering a more respectful exchange.

Here's a straightforward three-step structure that works well in most scenarios:

1. Observation: Describe a fact by focusing on a specific recent instance without interpretation or judgment.
2. Impact: Explain how that instance affects you personally, framing it from your perspective.
3. Request: Ask for a specific action you'd like the other person to take, focusing on what would help you or the team.

Illustrative Example

Imagine a team member feels frustrated because a colleague frequently interrupts them during meetings, causing them to lose their train of thought.

Instead of assuming the colleague's intentions, the team member could approach them one-on-one and structure the conversation as follows:

- Observation: "I've noticed that when I start sharing my ideas in meetings, you sometimes jump in with your own thoughts before I've finished."
- Impact: "When this happens, I feel flustered and find it difficult to communicate my ideas fully, which affects my confidence in sharing."
- Request: "I'd like to ask you to wait until I'm finished before jumping in."

Notice how this doesn't do the following:

- Assume intent—good or bad—on the part of the manager.
- Address or even presume to know what the manager was thinking.
- Ask the manager to explain themselves.
- Express judgment while explaining the situation.
- Suppress one's own feelings or emotions.

Again, everything you say is centered on your own experience and avoids imposing your assumptions on others—whether implicitly or explicitly—and using that against them.

The approach was developed to address conflicts that often arise from miscommunication and is effective in both personal and professional settings. This method for communicating noninvasively acknowledges interpersonal boundaries, personal integrity, and autonomy, while minimizing the risk of unintentional overreach.

This communication style is rooted in a framework known as Non-Violent Communication (NVC). For more information, visit cnvc.org.

Self-Advocacy Doesn't Have to Be Smarmy—It's Beneficial for Both You and Others

Self-advocacy and self-promotion often get a bad rap, especially in tech and cybersecurity, where many leaders shy away from it because they think "playing politics" isn't necessary.

But that is far from the reality.

In my experience working with cyber leaders and executive coaching clients, many believe in a meritocracy—that simply showing up, doing good work, and performing well will get them noticed, appreciated, and rewarded. But this assumption doesn't hold true, especially at the executive level.

Avoiding "politics" or resisting the idea of "selling" or "marketing" oneself—something many cyber leaders dislike—can result in missing opportunities to communicate the unique value you bring as a leader and executive:

- What you do, what you're good at, and why that matters.
- Your intent, goals, and personal beliefs, and how they fit into the bigger picture.
- Your experiences and strengths beyond cybersecurity.
- How you typically work, how others can best work with you, and where you're open to adapting.

By not sharing these aspects of yourself, you miss the opportunity to position yourself as a leader worthy of a full seat at the table. Self-advocacy is about clarity, communication, and ensuring that others understand what you bring to the organization.

Responsibly Self-Promoting

Self-promotion has two parts: your product and your user manual. Together, these help you communicate your value effectively.

1. **Product Description:** This is how you describe what you're good at—your strengths and unique contributions:

   a. "I'm good at engaging with customers and putting them at ease."
   b. "I'm skilled at motivating staff and generating loyalty."
   c. "I excel at _____."

Own this list of strengths. It's your product description.

2. **User's Manual:** People often joke about needing a manual for things like parenting or homeownership, but imagine if you provided one for working with you. Your user's manual can help others understand how to bring out your best, how to work effectively with you, and how to engage you to be as productive, efficient, and impactful as possible. It might include your preferred working style, whether you thrive with big-picture discussions or prefer getting into details. It could also outline what to avoid to prevent disruptions to your workflow. Here are some examples:

   a. "I prefer that the team handles logistics for group events, but I want a role in big meetings, even if it's just to kick off and close."

   b. "Bring me ideas early so I can help shape them" versus "I want ideas only after they're mostly fleshed out."

   c. "I don't need to see the final deck before you present" versus "I want final approval on the deck."

   d. "I execute better when I understand the project's importance to the company" versus "I need a clear sense of the final product to avoid getting bogged down in details."

Providing others with insight into your strengths, preferences, and needs helps them work with you effectively and builds stronger professional relationships.

Banishing Self-Discounting

Self-discounting most often occurs when receiving praise. Once praise is given, self-discounting can occur when you do the following:

- Change the subject.
- Diminish your role.
- Give so much credit to other team members that you ignore your contribution.
- Undervalue the principles, effort, skill, or creativity you brought to the work.

These behaviors, common among many professionals, especially women, undermine the recognition you've earned. Accepting praise can feel uncomfortable, but when someone offers it, they're recognizing your value. Deflecting praise can make the person who offered praise feel dismissed, and it raises questions:

- Do you not see the value you bring? If not, how will you know to intentionally leverage your attributes as strengths?
- Do you lack self-confidence? If so, should we limit your access to more challenging roles or tasks?

Neither of these is good for your career.

To counter this, consider an example:

Imagine your boss says, "That was the best leadership planning session we've run in years. You did a fantastic job, and everyone was really happy." Take a moment to think about how you'd respond.

Here's a deflection:

"It was no big deal. The material was already mostly developed, at least the key parts. We just had to practice a little." (Self-diminishing response.)

Here's an acceptance:

"Thank you. I think we really pulled it together. The team was great, and I helped by letting each person contribute in their area of strength, especially since it was such a high-visibility project." (Accepting response while giving due credit.)

Notice the difference? In the second response, "thank you" is a simple acknowledgment of the praise, followed by an explanation of your thinking, your role, and how the team contributed. This shows the person giving praise that you understand and appreciate it, inviting further feedback.

If accepting praise is an area you want to improve, try writing down instances where you received praise. Reflect on how you responded and consider how you could have accepted it more openly.

Raising Your Unique Contribution

Recognizing your unique contribution is another part of self-advocacy. Your path to success will be your own, shaped by your experiences and strengths. Reflect on the unique contributions you make to your team, projects, and organization. Acknowledge this uniqueness and embrace it—it's part of what makes you valuable.

Understanding Confidence Disparities in Self-Assessment

In studies around knowledge and confidence—where knowledge questions were asked combined with confidence questions about the respondent's answer (basically, how sure participants were of their answers)—the research found a huge gender disparity, with men tending to have confidence even when they performed poorly, and women having less confidence even when they performed well.

In another example, when reviewing a job description, women who don't meet 90 percent of the listed qualifications determine they are unqualified, while men who meet just 60 percent often believe they're qualified.

These dynamics frequently appear in the workplace, influencing small decisions, the risks we're willing to take, and the opportunities we pursue. These determine who applies for jobs, negotiates salaries, and gets promoted. Companies and leaders who overlook this risk creating a culture where confidence is mistaken for competence, exacerbating gender disparities. Leaders must adjust their evaluation processes to focus on ability and potential rather than self-promotion.

Where might this show up for you? For others? What can you adjust?

Accidental Diminishment of Others and Self

Accidental diminishment is a phenomenon where leaders unintentionally reduce the confidence, contributions, or effectiveness of others—despite their best intentions and often without awareness. Recognizing and avoiding accidental diminishment allows leaders to empower their teams and create a more supportive environment.

7.3 Don't Row if You Have a Sail; Don't Guess if You Have a Compass

Ambitious goals can help you break away from default patterns and stay on track toward meaningful outcomes. Intentional changes require envisioning a destination and mapping your future with purpose. Along the way, some discoveries serve as your compass—clear truths that guide your decisions.

Big audacious goals drive growth; easy goals soothe and slow.

Here's a question I pose to my clients:

"It's three years from now. We run into each other at Chicago O'Hare International Airport. You tell me you've had the best three years of your life. What have you done?" (Activities, not attainments.)

Write out your own responses to this question. I expect you to do this exercise because you made a commitment at the start of this section to truly invest in your goals.

You will, indeed, thank me later.

Take your time. I'll wait.

Many clients have answered, "I want to make tons of money."

It gets interesting and meaningful when I ask them, "What's the money for?"

One replied, "I had to put myself through college, and it was hell. I don't want to do that to my kids."

That's real, and that can motivate you to overcome many challenges at work. Don't underestimate the power of keeping that deep motivation in mind.

Another time I had a client say they wanted to retire early. After coyly avoiding the deeper question as to what they'd do after they retire, they eventually revealed they wanted to open a specific type of retail shop with their wife.

That's cool. That inspires me, and clearly it inspires them.

Go beyond the initial answer. Keep digging until you uncover something that relates to people you love or experiences you want—money and physical assets are just intermediaries. Go deeper:

RESPONSE (ATTAINMENT)	REFRAME (ACTIVITIES)
I made a ton of money.	What did you use it for?
I have a bigger title.	What does that empower you to do?
I have a huge house.	What does that offer you?
I am respected.	What does that mean to you personally?

Don't hide from your goals; dig into them.

Write them down, revisit them several times a year, and allow yourself to refine or even revamp them. This is about setting your compass toward the life you genuinely want.

Now that you've done that part, consider this question:

What's your unique edge? Review your personal assets—qualities that are uniquely valuable.

This is your personally unique edge, wherever you go, in the workplace or otherwise. Ask yourself what people say when they talk about how you contribute to the group. Ask what people say about how you contribute one-on-one.

This isn't about technical skills or table-stakes CISO strengths like "I break down complex topics" or "I'm a people-first leader." Every CISO needs to do that. In fact, if it's anything related to cybersecurity or a general management adage but can't be expanded to other parts of your life, you're likely on the wrong track.

Your edge points to something uniquely yours—a trait that always positions you more favorably in certain regards than anyone else in the room. An edge goes beyond strengths and values. You can google lists of strengths and values. An edge will be so unique to you that it won't show up on a list.

Once you find this edge, it tells you your most favorable terrain and therefore the path most suited to you. When your edge aligns you with the right opportunities, it accelerates your progress beyond what others might achieve. And it's almost always an advantage, if you know how to spot opportunities and use your edge to shorten the time or energy to get to the destination.

7.4 Break Free from Limiting Thoughts to See the Shackles You Wear

I've heard many complaints during my career; some are trite, some sound trite at first but aren't, and some are deep. And most are unsolvable as stated, but solvable in a fashion not being considered.

This comes down to habits of thought.

Catching someone with poorly framed complaints is a moment to pause, reflect, and ask challenging questions of the person who said it (or thought it, in the case of yourself).

This is especially true when you're the one doing it.

Without being cynical toward anyone, consider how familiar the following thoughts are to you? Perhaps from thinking in these terms, or in more "mature" variations. For clarity, these are stripped of subtlety:

1. This is really obvious; everyone should get it.
2. The business is being unwise. I'm tempted to give up on these people.
3. My hard work saved this company from so many potential breaches, so why don't I get credit?
4. I worked without sleep for a week when we were hit with a breach. I barely got a thanks.
5. No one sees how important cyber is.

6. Why don't people just listen?
7. They don't know what they're doing.
8. I know more than them.
9. This is foolish.

I hear these said less frequently than I used to. But the sentiments are echoed in a myriad of other phrases that boil down to the same thing.

Words may have been refined and softened, but in too many cases the deeper thinking hasn't evolved. Think about what thoughts frequently crop up for you about other people, and what that means for where you can take a step to validate your thinking, or make a change to your thinking to allow you to focus where you have the most control: yourself.

If you do quietly think these to yourself, I am challenging you. I'm challenging you to change the premises and the habits of your thinking. Ask yourself if you've thought any of these things recently.

Questions

1. What factors in my life have shaped how I react to challenges?
2. What personal experiences and history affect me? In what ways does this lead to future distress or dysfunction?
3. Am I satisfied with the support I receive to handle personal or professional stress?

CHAPTER 8:

Practicing with Today's Small Issues Prepares You for Tomorrow's Big Challenges

Sustainable change rarely happens from a single act or event. Our future story bends along an arc, with real growth emerging through consistent actions. By developing mechanisms that reinforce positive habits, you set yourself up for greater success as each action builds upon the last. These mechanisms can be especially powerful when used together to create steady, reliable progress toward your goals.

Habits can also help you in unexpected ways. A habit that you develop today, when things are relatively easier, builds muscles and ways of approaching problems that can pay off tenfold when you tackle the bigger challenges you will face tomorrow and beyond.

Recall my story of the "wrongest layoff"?

Think of this chapter as a guide to exploring, selecting, and practicing mental habits through practical techniques and tools.

Some of these techniques may become part of your daily routine, others may serve as training wheels, some may lead to self-discovery, and many may not fit your needs at all.

The true work of growth and impact is in experimenting with these tools to understand where they might support you, how to adapt them for personal and situational challenges, and how they can fuel long-term development. Avoid focusing on immediate results or quick wins, as these can overshadow the deeper value of the experience and limit your discovery of a potentially powerful tool.

The techniques and tools in this section will address several facets of the executive role:

1. Workplace communication
2. Professional stance
3. Self-advocacy
4. Sourcing insights

You'll benefit most by selecting across several types of technique—accountability, reflection, and systems—to win the battle between your current self and on a default path, and your desired future self.

8.1 Use Action Systems to Reduce Your Reliance on Willpower

Willpower can take you only so far. By setting up action systems, you reduce friction and reliance on sheer determination. These systems help you conserve energy by eliminating repetitive decision-making so you can focus on more important challenges.

Although these tools are often used in a corporate setting, I have found that the demands on an individual executive—be it a CISO or CEO—also rise to the level that the rigor and structure will be of great aid to the executive.

Goals and Milestones

Goal: Staying on the right path and maintaining focus.

Adages abound in this space (e.g., "If you don't have goals for yourself, someone else will"). As with many adages, they contain truth but not directions or instructions. We need tools and techniques to turn the sentiment into choices that lead to *right action*.

- Personalized Goal Setting: Set goals that are challenging yet motivating. While SMART goals (specific, measurable, achievable, relevant, and time-bound) are common, you might prefer frameworks like CLEAR (challenging, legal, environmentally sound, agreed, recorded), PURE (positively stated, understood, relevant, ethical), or HARD (heartfelt, animated, required, difficult) goals, which account for intrinsic motivation.
- Milestone Tracking: Break down large goals into smaller, manageable milestones, tracking progress to maintain momentum.
- Action Plans: Develop and share action plans based on feedback, ensuring transparency and accountability.
- Project Management Software: Use tools like Asana, Trello, or Jira to organize tasks, set deadlines, and track progress.
- Automated Reminders: Set up automated reminders for key tasks and deadlines to ensure nothing falls through the cracks.
- Dashboards: Create dashboards to visualize progress and key metrics, making it easier to stay on track.

- Goal-Oriented Planners: FranklinCovey, Bullet Journal Method, others.
- Celebrating Achievements: Recognize and celebrate milestones to sustain motivation and acknowledge your efforts.

Streamlined Decision-Making

Goal: Increasing time in zones of strength and value creation; reduce decision fatigue.

It is widely known that extremely busy but effective executives protect their energy and capacity for the most important things. This is done in several ways, from having a standard "uniform" (black suit and red tie for finance CEOs, black mock turtleneck for tech CEOs) to the more complex or extreme (telling the executive assistant to turn every email into a standing meeting agenda item or to delegate the email to a direct report).

These types of techniques are forcing functions for capitalizing on the executive's strengths and reducing time operating outside of that zone of strength. A forcing function is a design, mechanism, or process embed that intentionally creates a constraint, reminder, or obstacle to make sure a desired step occurs, or practice is avoided:

- Delegation: Delegate tasks effectively to empower your team and free up your time for strategic decisions.
- Decision Frameworks: Use decision-making and delegation frameworks like the Eisenhower Matrix to prioritize and make informed choices.
- Standard Operating Procedures: Develop and implement standard operating procedures for recurring tasks to streamline processes.
- Personal Principles: Establish core principles to guide consistent,

aligned decisions, especially in challenging or ambiguous situations.

- Decision Journal: Track major decisions and reasoning in a journal to learn from past choices and refine future decision-making.
- Habit Tracking: Regularly assess your progress on key habits to help you stay accountable and focused.

Eisenhower Matrix: Many people are familiar with variations of this concept, such as the simpler but less practical urgent-versus-important distinction. The Eisenhower Matrix (or Eisenhower Decision Matrix) is a time-management and prioritization tool often attributed to former U.S. President Dwight D. Eisenhower. It categorizes tasks based on their urgency and importance, guiding individuals to focus on what truly matters. Quadrant 1 (Urgent & Important) addresses crises and pressing deadlines that require immediate attention. Quadrant 2 (Not Urgent & Important) encompasses strategic activities like planning and skill-building, which should be scheduled for focused attention. Quadrant 3 (Urgent & Not Important) involves interruptions or tasks that can be delegated to others. Finally, Quadrant 4 (Not Urgent & Not Important) includes low-value activities and time-wasters that are best eliminated.

Resilience Strategies

Goal: Build mental and emotional resilience to sustain long-term effort.

Work is hard, and humans, from individual cells to organs to systems to body-level processes (mental and physical) almost always benefit from rest, and always have processes to repair themselves.

The most important aspect of blood pressure is the lower value; the most important part of training for a marathon is timing and degree of rest; building a muscle group in weight lifting happens only with two days' rest between lifting sessions. Research repeatedly shows that maintaining high levels of mental acuity requires both breaks and rest. We are replete with examples, and we work against ourselves when we don't concede this. The following practices engage the "rest and repair" part of our natural rhythm:

- Work-Life Harmony: Fully disengaging from work to prevent burnout and maintain high performance.
- Mindfulness Practices: Practices such as meditation or yoga to reduce stress and improve focus.
- Resilience Training: Programs to develop coping strategies and maintain a positive outlook.

Prompters

Goal: Use visual prompts and reminders to reinforce daily actions, helping you stay aligned with your long-term goals and values.

Relying solely on memory to keep your intentions and goals in mind adds an unnecessary layer of work on top of the tasks you already want to accomplish. Visual prompts serve as helpful reminders to keep your focus aligned with your intended actions and priorities.

- Visible Sticky Notes: Place key reminders on your monitor or desk with brief phrases that focus on daily priorities or motivational cues.
- Digital Dashboard Alerts: Set up dashboard prompts or screen savers that display your core values, goals, or daily focus.
- Phone Notifications: Use recurring reminders on your phone

to cue important habits, like taking a mindful pause, reviewing goals, or assessing progress.

- Physical Cues: Place objects in your workspace—such as a particular book, object, or card—that symbolize your commitment to specific goals or mindsets.

Life happens. Don't let the day-to-day make you forget that your goal is to retire in ten years to open a shop with your partner.

8.2 Seek Social Commitment for Accountability

Meeting the expectations of others can serve as a significant motivator, even when those expectations are self-imposed. When you commit to others to accomplish a task, the anticipation of fulfilling someone else's expectations—regardless of whether the task was self-assigned—exerts a form of social pressure.

Keeping commitments is an intrinsic human trait, and we are wired to not disappoint.

Several groups can help you leverage the psychological effect of self-imposed social pressure:

1. Circle of Advisors: Ask for accountability from the person who prompted you to rethink a situation or triggered a thought that you wanted to make a change. This serves you both, as they have insight into why you want accountability and can derive satisfaction from watching the idea they sparked turn into action.

2. Professionals / Paid Advisors: Hire experts whose financial interest is tied to your success, such as consultants or coaches who offer specialized insights and strategies. Their expertise

and vested interest in your progress enhance accountability, while the investment you've made in their guidance increases your own motivation to follow through.

3. Accountability Coach: Choose a professional coach rather than a boss or subordinate to maintain a balanced power dynamic. A coach's role is to support your growth, hold you accountable, and offer constructive feedback without bias, with regular sessions keeping you on track and helping to address any obstacles that arise.

4. Trusted Acquaintance: From your social circle, select a reliable individual who isn't a close friend but someone you can trust for honest feedback. This person can provide support and hold you accountable without the personal complications that might arise with closer relationships, giving you a healthy level of professional distance and trust.

5. Public Commitments: Make your commitments known within your organization or professional network. Publicly sharing your goals increases accountability by leveraging social pressure, as your progress (or lack thereof) is visible to others, which can boost motivation for those who respond well to external expectations.

8.3 Use Deep Thinking to Turn Experience into Meaning

Structured Reflection
Goal: Facilitate consistent growth and skill refinement.

- Daily Reflection Books: Use resources like *The Daily Stoic* for structured daily reflections.

- Weekly Reflection Sessions: Set aside time each week to reflect on what went well, what didn't, and how to improve.
- Prompted Reflection: Use specific prompts to guide your journaling, such as "What did I learn today?" or "How did I handle a difficult situation?"

Decision Journaling

Goal: Enhance decision-making skills and learn from past experiences.

- Daily Entries: Spend ten minutes each day writing about key decisions, including context, options, rationale, and expected outcomes.
- Weekly Review: Reflect on the decisions made during the week to identify patterns and insights.
- Mentor Feedback: Share selected entries with a mentor or coach during check-ins for external perspectives and advice about your process, thinking, and criteria (not the decision itself).

Structured Journaling

Goal: Capture insights and track personal development over time.

- Daily Journaling: Spend ten to fifteen minutes each day writing about your experiences, thoughts, and feelings related to your leadership journey.
- Leadership Journaling: Keep a leadership journal to document experiences, insights, and lessons learned.
- Expectation Journaling: Record your expectations—of yourself, others, and situations—and reflect on whether they were met and adjust as warranted.

- Outcome Tracking: Regularly compare expected outcomes with actual results to assess and improve decision-making accuracy.

Win Tracking

- Success Log: Maintain a log of successes to recognize achievements and reinforce progress.
- Gratitude Log: Things you are grateful for to cultivate a positive mindset and increase creativity and innovation.

After–Action Reviews

Goal: Learn from recent events to improve future performance.

You can try to do this on your own, but you are likely best served by engaging your circle of advisors to help you with the following:

- Post-Project Analysis: After completing significant professional projects, conduct a review session to discuss what went well, what didn't, and what can be improved.
- Event Debriefs: Conduct debriefs after major career or professional events to analyze your actions, responses, and outcomes.
- After-Action Reviews: Conduct after-action reviews following significant projects or events to capture lessons and plan for future improvements.

Mindfulness and Meditation

Goal: Enhance self-awareness and reduce stress.

- Daily Meditation Practice: Incorporate a daily meditation practice to clear your mind and reflect on your thoughts and emotions.

- Mindfulness Techniques: Use mindfulness techniques throughout the day to stay present and aware of your reactions and behaviors.
- Reflective Walks: Take regular reflective walks without distractions to think deeply about your leadership journey and challenges.
- Scheduled Meditation: We discover things about ourselves with space and in silence. Someone who is always on the go is not likely to grow.

Learning and Development

Goal: Continuously expand your knowledge and skills.

- Professional Development Plans: Create and regularly update a professional development plan outlining your learning objectives and strategies. Be conscientious about the balance of focus between developing as an executive, what's comfortable, and what is interesting.
- Continuous Learning: Dedicate time each week to reading industry literature, taking online courses, or engaging in other learning activities.
- Feedback Integration: Incorporate insights from reviews into ongoing techniques such as action plans, principles, and habit development.

Regular Self-Assessment

Goal: Cultivate self-awareness and highlight areas for improvement.

- Weekly Self-Review: Dedicate thirty minutes each week to

assess your performance, noting successes, challenges, and areas for growth.

- Monthly Progress Check: Conduct a more comprehensive review at the end of each month to evaluate your progress toward your goals and adjust your strategies as needed.
- Quarterly Reflection: Set aside time every quarter to reflect on your broader career goals, personal growth, and alignment with organizational objectives.

Incorporate these incrementally. Explore what works for you without feeling the need to commit.

COMMON "WHEN TO USE" SITUATIONS	TOOLS
When seeking to process experiences or emotions.	Leadership Journaling, Daily Journaling, Prompted Reflection, Reflective Walks
When aiming to build a habit of self-reflection.	Expectation Journaling, Weekly Reflection Sessions, Daily Journaling
When needing to monitor progress and adapt strategies.	Weekly Reflection Sessions, Weekly Self-Review, Monthly Progress Check
When needing to align personal growth with broader goals.	Quarterly Reflection, Professional Development Plans, Monthly Progress Check
When seeking to cultivate self-awareness and emotional regulation.	Daily Meditation Practice, Mindfulness Techniques, Expectation Journaling
When aiming to stay informed and enhance knowledge.	Continuous Learning, Professional Development Plans, Workshops and Seminars

When trying to track emotional and mental patterns.	Daily Journaling, Success and Gratitude Log, Expectation Journaling, Decision Journaling, Assumption Journaling
When needing to stimulate thinking or process complex issues.	Reflective Walks, After-Action Reviews, Prompted Reflection

As you experiment, check in with yourself on how you're doing with them, and whether they are working for you. Be real with yourself and change or abandon what's not working if it doesn't seem like it will work for you.

Specifically, consider these common "when not to use" instances:

- When facing immediate and urgent decisions that require quick action.
- When time constraints or external conditions make the practice impractical.
- When it becomes a chore rather than a meaningful activity.
- When it feels forced, superficial, or insincere.
- When external factors necessitate more frequent adjustments than the practice allows.
- When lacking the time, space, or environment conducive to the activity.
- When they lead to overanalysis or cause hesitation or self-doubt.
- When the practice starts feeling repetitive or mechanical, reducing its effectiveness.

In general, if you are under stress, try not to add to your stress by taking on these activities in big ways unless you're sure it's going to help. Instead, consider using any of these once, without further commitment, as a method to address your immediate situation. If it helps, that may be a signal to expand your testing, or to reattempt it at a later time.

Questions

1. How do you distinguish between being aware of a tool or technique and effectively applying it in your work?
2. What are common habits of thought that you recognize in yourself that may need reevaluation?
3. What thinking and meta-cognition muscles can you build now to make decisions, relationships, and strategy easier in the future?

CHAPTER 9:

Sustainable Growth Thrives on Steady Mechanisms

9.1 Mindset First, Solutions Follow

When I coach my executive clients, I intentionally avoid offering typical cybersecurity approaches or tactics—no trite advice, clichés, or even solutions that have proved successful for others or myself. Why? Because this kind of advice is often based on several flawed assumptions:

- The advice giver fully understands the unique social, political, relational, and psychological dynamics at play.
- The recipient is ready to take action and simply lacks knowledge of what steps to take, with the advice giver assuming the receiver has no internal obstacles holding them back.
- The circumstances are sufficiently similar, or the advice giver

has a deep enough understanding to provide sound guidance
from their experience.

- The receiver will grasp the nuances, implications, challenges,
  and success factors associated with the advice.
- The offered solutions align with the receiver's values, principles,
  and priorities.

These flawed assumptions are also why even the best management
consultants cannot provide ready-made solutions to address such complexities. Truly customized consulting solutions require significant
investments of time and money to navigate the unique challenges of
each situation.

Here's an example in action.

We all hold beliefs about what others think of us, but too often
these beliefs work against us. Cybersecurity executives, given the
unique demands of our roles, feel this acutely—especially when we
start assuming how other business leaders view us.

A common sentiment among clients has been some variation of
"This person feels negatively toward me." This is a frequent human
reaction, something most of us have experienced at one time or another.

Variations include the following:

1. This person doesn't like me.
2. They don't care for me or my personality.
3. They don't respect me.
4. They don't care about my success / about cyber / about risks
   to the company.

Feelings and instincts are signals, and what we do with them determines our choices between a path to growth and confidence versus a path to continued or increased friction.

One question to ask yourself is simply *What if that's not the case? What if the advice you get solves the wrong problem?*

Through the process of coaching, most clients realize for themselves their own assumptions, dispense with them, and want to plan to build a relationship or take action that moves them forward. In every case, the client feels a weight lifted and is energized and ready to rethink and reconfigure their mental, if not actual, relationship with this person. This high-friction point is gone and no longer creates the stress it had been when we started the conversation.

It's amazing to see the impact.

Again, if you ever think something or someone is working against you, ask yourself, *What if that's not the case?*

9.2 Use Self-Coaching to Best Leverage Your Own Resources

Coaching is a demanding craft, requiring deep intention and careful focus, and it is fraught with traps. Self-coaching is even harder. For this reason, even seasoned coaches have their own executive coaches.

While it's not impossible, and while self-coaching will rarely be as robust as having a professional coach, you can still derive benefits from employing self-coaching when a professional coach is financially out of reach for your company—or for you personally.

How does one coach oneself? Following a process similar to coaching can be helpful:

1. Create space: Designate a physically undistracting place and

specific times to which you are committed to the degree of a formal obligation to someone else (*ahem*: your future self).

2. Be present to yourself: Set aside your task lists and turn down your self-imposed pressure on time and attention. Turn off alerts and notifications.

3. Get in the mindset: Open yourself to creative ideas and your own thoughts, feelings, and instincts—and, if appropriate, critical self-evaluation. Watch a comedian or a dad joke video for a few minutes. Humor is implicit permission to play, and it opens up creative space. Laughter offers a biochemical wash that reduces self-critical thinking.

4. Brainstorm several potential topics: Brainstorm two or more topics, then pick the topic more important or pressing to you.

5. Ask yourself what you can tackle in the time you've allotted to make progress.

6. Ask yourself what meaning the topic has for you. Generally the first thing you think of is at or near the surface. Seek to go beyond that. This is a good place to practice the "five whys" technique.

7. Ask yourself what success looks like: If this issue was resolved, what would that look like?

8. Identify why this matters to you: Connecting activities to personal meaning and purpose fuels motivation and helps overcome challenges more effectively than willpower alone. This linchpin question in coaching should not be overlooked, as meaning-driven actions sustain commitment and progress.

9. With this foundation, follow your instinct. Consider talking out loud. Hearing yourself talk things out often opens insights.

10. Reflect and commit.

While it isn't always obvious in coaching, most of the above is the setup, and as with many things, setup is important. An experienced coach will know how to vary or sometimes appear to skip steps. But as a beginner coach, or in self-coaching, you follow the full practice. It's exceedingly rare to have a good coaching session without a good setup.

Real, viable solutions come from deeply understanding the core of the problem by applying the right, most meaningful question for the moment, considering the individuals involved and the broader context. This process demands significant tacit knowledge, as well as the ability to make choices that both complement and challenge strengths. It also requires drawing from experiences while setting aside limiting self-beliefs.

That said, some universal coaching concepts can be helpful and applied repeatedly. Once learned, these questions will serve you throughout your career, and likely in life as well. Deep practice with these questions is part of what clients gain from coaching in the long term.

In this book, I have shared key mindsets and tools with you. But as with most things human, we are most bought into and most likely to adopt what we know to be true. Truth comes from ourselves much of the time.

With that in mind, I've shared practical ways (below) to tap into yourself to meet your own challenges. These questions will help you draw more deeply from your own experiences, positive self-identity, and determination to achieve future success. Use them when facing challenges or difficult decisions.

Self-Reference

- When have I faced a similar situation in the past?
- What's worked for me in situations like this before?

Insight Sourcing

- What philosophy did someone I admire use when faced with a similar challenge?
- What advice would I give someone else in this situation?

Time Shifting

- What would my future, ideal self do?
- What advice would my future self give me?

Personal Assumption Checking

- I think someone has a view counter to my goals. Have I objectively verified it? What can I do to validate this?
- Am I sure this person's behavior is about me, or could it be something they're dealing with?

Situational Assumption Checking

- What, if anything, is truly unique about this situation that I haven't faced before?
- Is it likely someone else has faced this before? What might they have done?
- What if something I believe about the situation isn't true?

Inversion

- What's the worst possible outcome? How can I avoid the worst possible outcome?

- Do my default habits and behaviors lead me to a place I don't want to go? How do I change them?

Blue–Sky Thinking

- If there were a solution to this problem, what would it be?
- If I could change one thing about this situation, what new possibilities emerge?

Hesitancy Release

- If I could do anything, without constraints or worrying about consequences, what would I do?
- What's the worst that could happen if I did what I felt best in the situation?

I encourage you to keep this handy and come back to it occasionally.

9.3 You Are Your Most Important Partner

There are many ways to grow. The key is to find what works best for you. Choose the methods, approaches, and tools that align with your personality, strengths, situation, needs, and goals. Different approaches work for different people, so experiment to find what supports your growth and keeps you on track.

Questions

1. How honest am I about what's not working in my current approach, and what would it take for me to adjust or let go of ineffective approaches?
2. In what ways do I resist or embrace feedback as a tool for growth? How can I become more open to receiving and acting on it?
3. What positive role can I play as a coach to myself?

Conclusion

The role of the chief information security officer is no longer just about managing threats—it's about leading across the organization. As business paradigms evolve and volatility and uncertainty increase, the stakes have never been higher, and the challenges have never been more complex. With the right mindset, tools, and strategies, CISOs can turn these challenges into opportunities for growth, resilience, and innovation.

This book has aimed to equip you not only with practical insights but also with a new perspective on what it means to be a CISO. The future of cybersecurity leadership isn't about fighting battles alone; it's about fostering collaboration, changing the way we think about our calling, and stepping into the role of being a member of the CEO's team. As CISOs, we can elevate our role to that of a general advisor in an increasingly uncertain, unpredictable, and adversarial world.

The role of the cybersecurity executive is far from settled. We are in a period of evolution, revolution, and transformation. The rules are being written, and we have the opportunity to help write them.

If we don't seize this moment, other roles will fill that void, and we will miss our chance to offer our unique skills and insights.

How you position yourself as an executive unlocks your impact, which has a profound influence on your organization, your team, and society at large. By embracing a broader perspective and the skills you've developed, you can help shape a better world for your company, community, and country. A safer, more secure future for everyone.

Our field has an opportunity to give birth to another visionary like Dorothy Denning, Eugene Spafford, Alan Paller, or Wendy Nather.

I hope that person is you.

Appendix

A: Questions to Move from Concepts to Action

Questions to Assess

- Do I rely on one method for personal change, or do I thoughtfully integrate mechanisms like action systems, social commitment, and reflection?

- What might I gain by diversifying my approach to personal growth? In what ways could each of these mechanisms add value to my development journey?

- How much do I rely on willpower alone to get things done? Could better systems help me achieve my goals with more ease and less strain?

- What types of systems could I introduce to reduce reliance on willpower? How might these systems free up mental energy for other priorities?

- How well do I incorporate feedback into my growth journey, and how could I create stronger systems to track and act on it consistently?

- In what ways could a feedback tracking system deepen my self-awareness and accountability? How might I ensure that I apply insights consistently to foster improvement?

- Do I make enough space for intentional reflection? How might more structured reflection practices support my personal and professional growth?

- What specific reflection practices could help me focus on progress over time? How might regular, structured reflection guide me in adapting to new challenges more effectively?

- When challenges arise, do I address them primarily through my existing strengths, or am I willing to adopt new approaches that stretch my abilities?

- What might I gain by intentionally cultivating skills outside my comfort zone?

- How do I define success for myself, and does this definition allow for both short-term achievements and long-term growth?

- What adjustments could help me feel more fulfilled in both areas?

- When setting goals, do I focus solely on outcomes, or do I also pay attention to the process and the quality of my effort?

- How could a focus on process enhance my resilience and adaptability?

- Am I consistent in creating opportunities to solicit feedback, not only from those who know me well but also from diverse perspectives?

- What insights might I uncover by regularly seeking input from people in different roles or fields?

- In what ways do I foster curiosity in my everyday life and work?

- How might cultivating a mindset of curiosity improve my adaptability and openness to change?

- Do I actively prioritize mental and emotional resilience as part of my growth journey?

- What practices or habits could I integrate to support a balanced approach to personal and professional demands?

- How effectively do I distinguish between tasks that require precision and those where progress is more important than perfection?

- What impact could a greater focus on progress have on my overall productivity and satisfaction?

- When my priorities shift, do I readily reassess my systems and habits to ensure they remain aligned?

- How might regular reassessment help me stay aligned with my evolving goals?

- Do I engage in enough forward-looking practices, such as visualizing possible challenges and outcomes, to enhance my preparedness?

- How could these practices strengthen my confidence in making decisions under uncertainty?

- Am I mindful of the habits and mindsets I unconsciously carry over from past experiences?

- Which of these might serve me well now, and which ones could benefit from thoughtful revision?

Questions for Action and Implementation

- What methods of self-coaching and accountability best fit my work style and personality? Have I experimented enough to discover what truly works for me?

- How might trying new methods deepen my understanding

of what keeps me motivated? Which small adjustments could help refine my current self-coaching approach?

- Which mechanisms (e.g., action systems, social commitment, reflection) align most with my current stage of development, and how can I better integrate them into my daily routine?

- What would a stronger daily routine look like with these mechanisms fully in place? How might I measure the impact of integrating each one?

- What practices would help me stay more mindful and connected to my purpose, allowing me to approach my actions with greater intentionality?

- How could I build these mindful practices into my day naturally? What cues or reminders would help me return to my purpose when I get off track?

- How can I set goals that align deeply with my personal motivations, moving beyond frameworks like SMART and embracing a more intrinsic approach?

- What would an intrinsically motivated goal look like for me? How might aligning my goals with my values and passions enhance my sense of fulfillment?

- Who in my life can I trust for honest, consistent accountability, and how can I strengthen and leverage those connections to sustain long-term growth?

- How could I foster deeper relationships with those who offer honest feedback? In what ways might I reciprocate their support to create mutual accountability?

- Which key people in my resource network can I lean on for feedback and fresh ideas? How can I better cultivate and maintain those relationships?

- What specific actions would strengthen my connections with these individuals? How might I show appreciation for their support to encourage an ongoing exchange of ideas?

B: The Real Budget Process

An often-overlooked aspect of budgeting is the influence you can exert beyond simply drafting and submitting your own budget request. There are ways to support cybersecurity resource allocation across the organization without directly controlling every dollar. Effective budgeting doesn't always require owning large portions of the budget; it often depends on strategic influence and resource alignment.

- Understanding Stakeholders
  - o Timeline: Ongoing; validate at T-minus 16–20 weeks.
  - o Goals: Identify the key stakeholders in the budget decision-making process. This includes those who are accountable for the budget, those affected by the process, data stakeholders, and those responsible for implementation.
  - o Checklist:
    - ☐ Identify all stakeholders in the budgeting process (accountable, impacted, and implementation roles).
    - ☐ Map out each stakeholder's interests, priorities, and influence level.

    ☐ Begin initial outreach to understand their perspectives on the budget needs.

- Understanding the Official Process
  - Timeline: Ongoing; validate at T-minus 14–18 weeks
  - Goals: Familiarize yourself with the formal budgeting models established by finance and budgeting teams, including any budget coordination or consolidation functions managed by your superiors.
  - Checklist:
    - ☐ Review formal budgeting models provided by finance or budgeting teams.
    - ☐ Identify any coordination or consolidation responsibilities held by your boss or department.
    - ☐ Clarify timelines, submission requirements, and any mandated formats or documentation.

- Understanding the Parallel/Political Process
  - Timeline: Ongoing; validate at T-minus 12–16 weeks.
  - Goals: Recognize the informal and political aspects of the budget process, such as internal dynamics and unwritten rules that can influence decisions. This knowledge is often as crucial as understanding the formal process.
  - Checklist:
    - ☐ Identify informal power dynamics, alliances, and potential influence points in the budgeting process.
    - ☐ Talk to colleagues with experience navigating the political landscape for additional insights.
    - ☐ Document any unwritten rules or considerations that could affect proposal approval.

- Building Relationships with Decision-Makers
  - Timeline: Ongoing; rewarm as needed at T-minus 10–14 weeks.
  - Goals: Engage with well-informed advisors to gain insight into the perspectives and expectations of key stakeholders. Develop and refine your proposal by seeking input and feedback from trusted allies and likely advocates before approaching primary decision-makers.
  - Checklist:
    - ☐ Reach out to informed advisors for guidance on decision-makers' perspectives.
    - ☐ Identify potential allies and advocates within key departments.
    - ☐ Obtain preliminary feedback from trusted stakeholders to refine your initial proposal draft.

- Framing Needs and Value with Decision-Makers
  - Timeline: T-minus 8–12 weeks.
  - Goals: Ensure that your proposal speaks to the specific needs and priorities of the business. Use business language and align your proposal with broader organizational goals. Build credibility by showing that the proposal has been reviewed and supported by relevant departments, such as finance and procurement.
  - Checklist:
    - ☐ Tailor your proposal to address specific business needs, using clear business language.
    - ☐ Align the proposal with the company's larger goals and strategic priorities.

☐ Secure initial support or endorsement from relevant departments, like finance or procurement.

- Iterating and Refining the Proposal
  o Timeline: T-minus 4–8 weeks.
  o Goals: After initial discussions, gather feedback and make adjustments to improve the proposal's chances of approval. Maintain ongoing engagement with stakeholders to secure their support and address any emerging concerns.
  o Checklist:
    ☐ Present the initial proposal draft to select stakeholders for detailed feedback.
    ☐ Adjust and refine the proposal based on feedback to enhance clarity and alignment.
    ☐ Continue engaging with stakeholders, confirming support and addressing any new concerns.

- Presenting the Proposal
  o Timeline: T-minus 2–4 weeks.
  o Goals: Present your proposal in a way that emphasizes its value to the business while minimizing its impact on other departments. Be prepared to address potential objections, providing clear explanations and justifications.
  o Checklist:
    ☐ Prepare a clear, concise presentation of the proposal's business value.
    ☐ Anticipate potential objections and prepare responses with specific justifications.
    ☐ Emphasize how the proposal aligns with organizational priorities and minimizes disruptions to other departments.

- Submitting the Budget Request
  - Timeline: T-minus one week.
  - Goals: This final step may seem straightforward, but the groundwork above is essential to ensure a smoother approval process.
  - Checklist:
    - ☐ Ensure all supporting documentation and endorsements are in place.
    - ☐ Double-check that the final proposal meets all formal submission requirements.
    - ☐ Submit the budget request, following up with stakeholders to confirm receipt and reinforce support.

You may already be familiar with some of these methods, and you may have formed opinions based on your personal experiences. However, revisiting these steps offers fresh perspectives and helps adapt these strategies to your current situation, further supporting your growth as an executive.

C: Involving Others in Security Responsibilities

An effective security budget strategy goes beyond isolated requests. By approaching the budget process as a way to influence and integrate security responsibilities across projects, departments, and daily operations, you can foster a shared commitment to cybersecurity throughout the organization. This appendix outlines strategies to approach security budgeting and responsibilities from different angles, encouraging collaboration and sustainable investment.

Transactional Security Funding

How can I secure a one-time security budget increase within specific projects or procurements?

- Milestone-Triggered Funding: Establish funding releases tied to project milestones that include security assessments and implementations, ensuring security is addressed throughout the project life cycle.

- Bundled Procurement Packages: Integrate security as a mandatory part of procurement packages, so each project begins with allocated security funding.
- Conditional Approval Processes: Use conditional project approval processes to require a baseline security budget, allowing projects to proceed only if security costs are covered.
- Automated Allocation Rules: Automate a fixed percentage allocation of each new project's budget to security, standardizing security funding across projects.
- Exposure-Adjusted Budgeting: Tie additional budget allocations to the degree that business units or departments create additional exposure for the organization (recover department-level externalities).

Scaled Security Investments

How can we ensure our security investments grow in proportion to organizational expansion and complexity?

- Base + Count: Use a hybrid budget model with a fixed base for core security services plus additional funding based on specific metrics (e.g., endpoint count for endpoint detection and response [EDR]).
- Modular Budgeting: Create a modular budget approach that funds core security needs, with additional funds based on project-specific risks and requirements.
- Exposure-Based Cost Allocation: Allocate budgets according to the risk profile of each department or project, channeling more funds to high-risk areas. This approach aligns spending with actual risk and highlights areas with inflated risk but minimal business value.

- Organizational Risk Tiers: Establish tiers for security funding based on the sensitivity of data or processes within each department. Critical departments or high-risk projects would automatically qualify for a higher security budget, while lower-risk areas receive a basic allocation.

Operational Security Integration

How can we seamlessly integrate security into daily operations and collaborate cross-departmentally to strengthen resilience without disrupting workflows?

- Cross-Department Initiatives: Launch joint initiatives, such as cross-training between development and security teams, to encourage shared responsibility for security. Jointly present the case for funding for these initiatives.
- Allocation of Points in Coding Sprints: Allocate points for security tasks in coding sprints, allowing security to be addressed alongside feature development. (Note: This requires careful oversight to prevent overcommitment or misallocation by teams.)
- Integrate Security into Existing Processes: Continuously identify ways to make security less operationally distinct by embedding it into existing practices and processes so it becomes a natural part of daily workflows across departments.
- Embedded Security Roles: Place dedicated security liaisons within key departments (e.g., development, IT, HR) to ensure continuous security oversight. These embedded roles can advocate for security needs, address issues as they arise, and strengthen cross-departmental alignment.
- Security as a Key Performance Indicator (KPI) in Operational

and Personnel Reviews: Include security metrics as part of regular operational reviews or performance evaluations, encouraging departments to consider security as a key factor in their day-to-day responsibilities.

- Departmental Security Champions: Create a security champion program where select individuals in each department receive training on security best practices and advocate for security measures in their teams. Champions can act as a bridge between the security team and other departments, fostering a culture of shared responsibility.

Distributed Security Responsibilities

How can security responsibilities and budget considerations be shared across departments?

- Position as Business as Normal: Normalize cybersecurity as a routine part of business operations, like ensuring quality standards or project success, helping it become a nonnegotiable budget item.
- Transparent Security Reporting: Create regular reports on security posture, including department-specific metrics, and share them widely across the organization. Transparency encourages departments to take ownership of their security performance and fosters a collective understanding of organizational risks.
- Localized Security Budgets: Allocate small, discretionary security budgets to each department, allowing them to address minor security needs independently. This creates a sense of ownership and responsibility for security within each department, while larger initiatives remain centrally managed.
- Business Case Development Workshops: Conduct workshops

that help departments articulate their security needs in the language of business value. Departments gain tools to present security requirements in ways that align with the organization's goals, reinforcing cybersecurity as a core aspect of business success.

D: Contributing to Group Discussions on Topics You Are Unfamiliar With

The Six Thinking Hats is a concept developed by Edward de Bono, first introduced in his book *Six Thinking Hats*, published in 1985. De Bono's method encourages different modes of thinking by metaphorically "wearing" one of six colored hats, each representing a distinct perspective. This framework is widely used to facilitate structured group discussions and enhance decision-making by examining issues from multiple angles.

As a cybersecurity executive, you can enhance group dynamics by "wearing" a hat that isn't represented in a meeting. Doing so enables you to bring fresh perspectives or insights that may help tackle the problem more effectively.

This framework uses six "thinking hats" to guide contributions:

1. BLUE: Big Picture & Process
2. WHITE: Facts & Information

3. RED: Feelings & Emotions
4. BLACK: Risks & Downsides
5. YELLOW: Benefits & Positives
6. GREEN: New Ideas & Innovation

Below are sample questions for each hat. These questions are examples that work for me, but to make them effective, they need to be similar to *your* style of thinking and in *your* voice. Adapt them, or create your own, so that you're comfortable asking them, can offer clarification if needed, and can ask relevant follow-up questions.

Consider these samples as starting points for developing questions that align with each "hat."

Sample BLUE / Big Picture & Process Questions

- Which option best aligns with how we see ourselves as a company?
- What would the (possibly deceased) founder(s) say about this?
- Do we have the right people in the room for this discussion? Should we invite _____?
- Are we getting ahead of ourselves? Should we revisit the previous step and explore it more deeply?

Sample WHITE / Facts & Information Questions

- What kind of data would help us feel more confident in making a decision? *(Follow-up: How can we get this data?)*
- Are we sure that the data presented is the most relevant for this decision?
- What was the original mandate and intent of this initiative?

- What parameters or limitations are we working within? Are these actual constraints or assumptions?

Sample RED / Feelings & Emotions Questions

- How is everyone feeling about the direction this is taking?
- What emotions are coming up for each of you in these discussions?
- What's your overall sense of this topic and our approach to it?

Sample BLACK / Risks & Downsides Questions

- What potential risks or downsides might we have overlooked?
- Could there be any unintended consequences of this decision?
- What could go wrong if we proceed with this option?
- Are there any warning signs or red flags we should pay attention to?

Sample YELLOW / Benefits & Positives Questions

- What are the benefits if we move forward with this?
- How could this decision create new opportunities for us?
- What strengths or advantages do we have that will support our success here?
- In what ways could this enhance our company's mission or vision?

Sample GREEN / New Ideas & Innovation Questions

- What are some alternative solutions we haven't explored yet?

- How can we approach this problem in a completely new way?
- What innovative, risky, or unconventional approaches could we try to achieve our goals?
- Are there any outside-the-box ideas that might be worth considering?

Use these examples as a foundation to create your own questions, tailored to your style and the context of your discussions.

E: Learning About Your Peers Through Associations

Consider joining or guest-participating in the following associations to gain insight into executive roles:

Boards of Directors
- National Association of Corporate Directors (NACD)

Management
- American Management Association (AMA)

Human Resources and Training
- Society for Human Resource Management (SHRM)
- Association for Talent Development (ATD)

Risk Management
- Risk Management Association (RMA)
- Society of Information Risk Analysts (SiRA)
- Risk and Insurance Management Society (RIMS)

Finance and Accounting

- Financial Executives International (FEI): A global organization for senior finance executives.
- Institute of Internal Auditors (IIA): A professional association for internal auditors, providing guidance and certification.
- Institute of Management Accountants (IMA): An association for accountants in management.

Marketing and Sales

- American Marketing Association (AMA): A global professional association for marketers.
- Sales Enablement Society (SES): A community for sales enablement professionals.
- Customer Success Association (CSA): An organization for customer success professionals.

Technology and Innovation

- Association of Information Technology Professionals (AITP): A professional organization for IT professionals.
- Project Management Institute (PMI): A global professional membership association for project managers.
- Society for Information Management (SIM): A professional association for senior IT executives.

General Management

- American Management Association (AMA): A global professional association for managers.
- Executive Women International (EWI): A global organization for women in leadership.

Governance and Compliance

- Association of Corporate Counsel (ACC): A global bar association for in-house counsel.
- Society of Corporate Compliance & Ethics (SCCE): An organization for corporate compliance professionals.
- Society of Corporate Secretaries and Governance Professionals (SCSGP): A professional association for corporate secretaries and governance professionals.

These associations provide a range of resources, including networking opportunities, conferences, certifications, and publications to support executives in their roles.

Note: You'll notice that no CEO organizations are included here—such groups are often highly selective and focused on exclusive, commercially driven networks. Accessing these groups can be challenging until you're operating at a peer level within the C-suite.

F: Learning About Your Peers Through Relationships

External advisors and connections with cross-domain executives are essential for CISOs to broaden their perspective. The following are prime candidates for a CISO for cross-domain executives:

- Chief Executive Officer (CEO)
- Chief Financial Officer (CFO)
- Chief Operating Officer (COO)
- Chief Technology Officer (CTO)
- Chief Marketing Officer (CMO)
- Chief Human Resources Officer (CHRO)
- Chief Legal Officer (CLO) / General Counsel
- Chief Information Officer (CIO)
- Chief Strategy Officer (CSO)
- Chief Risk Officer (CRO)
- Chief Compliance Officer (CCO)
- Chief Diversity Officer (CDO)
- Chief Sustainability Officer (CSO)

- Chief Product Officer (CPO)

G: Developing Your Organization's Board of Directors

The approach of going to the board to support the cybersecurity budget or make decisions on specific issues mistakenly treats the board as an integral part of organizational operations, which is fundamentally flawed. Don't do it.

To reiterate the point another way, the board, lacking operational responsibility, should not dictate the organization's day-to-day structure, practices, or decisions.

By assuming they should overlook the CEO's role as the primary liaison between the board and the organization and the CEO's role as the key architect of the organization, CISO think it wise to leverage the board for CISO ends. This is a mistake. Don't use their attention to drive behavior or decisions.

The traditional role of a board of directors was seen as fiscal watchdogs and stewards of the company, focusing on oversight. This has changed. Board members are more engaged, are more informed, and offer more insights than in the past.

However, the modern approach sees boards as actively helping the company achieve its goals, engaging in discussions beyond financial oversight. The primary responsibility of the board has never been to make operational decisions or interfere in day-to-day management but to assess the CEO's judgment and provide support to the CEO to execute the company's strategy. This understanding is crucial when seeking to engage with the board, as both the CEO's and your performance are evaluated in this context.

Don't Misunderstand the Role of the Board of Directors

Today, boards have two primary responsibilities:

1. Hiring, evaluating, and potentially replacing the CEO.
2. Ensuring the company has a viable strategy and is executing it.

For the first responsibility, evaluation primarily involves assessing the CEO's judgment, including decisions related to appointing you as the CISO, and your performance.

Beyond these roles, commercial boards do not directly make operational decisions, such as budget allocations for cybersecurity. They focus on listening, observing, offering insights, and challenging proposals, while keeping their involvement at an advisory level to support the CEO in leading the company to success.

Your CEO Has a Relationship with the Board; Your Relationship with the Board Is as a Proxy for the CEO

Your CEO plays a central role in your ability to engage with the board. Understanding the perspectives and functions of both the CEO and the board can demonstrate readiness to engage with the board and facilitate a more compelling case for increased engagement with the board, but only if you work with the CEO's agenda.

CEOs have distinct styles and approaches. Some prefer to lead discussions, inviting others as needed based on specific circumstances or at the board's request. If your CEO operates in this manner, it presents an additional challenge for you to regularly engage with the board.

If you need to—and this will resonate if you enjoy sci-fi—consider the CEO as a point of interdimensional gateway where the rules of physics are different on either side. CEOs tighten the portal at times to protect both worlds.

A Note About Board Prioritization and Support

The best way to get the board's attention is to work with members of the senior executive team individually to collectively come up with the set of threats that have a cyber vector or initiation point but that ultimately accrue to threats to the business that the business is already worried about—be it fraud, or loss of revenue, or supply chain issues—and then *let those executives drive the conversation on how it could happen via cyber.*

Understand the CEO's Take on Your Job Before Engaging with the Board of Directors

Your first job as a CISO working with the board of directors is to not disrupt a functioning board, or make an underfunctioning board worse. You won't make friends with the CEO if you do, and you might damage the company in ways more significant than a cyber breach could.

Understand the Current State of Board Effectiveness

Half of the CEOs of the Fortune 500 companies think their board processes are ineffective. It's not easy to show up and make things happen at the board level, even for the CEO. Right out of the gate, most CEOs have challenges with the board. Appropriately, fixing that is their most important priority.

If you exhibit signs that you're going to disrupt the most strategic work of the board, how eager is the CEO going to be to invite you?

Be on the CEO's team first in order to get the invitation to engage the board

Your opportunity to present to the board frequently depends on the CEO's agenda. As you consider strategies for engaging with the board and obtaining permission to do so, be mindful of factors that influence the CEO's decision-making.

Understanding the roles and the dynamic within the board, who collectively serve as the CEO's evaluators, is a significant and important factor.

Boards are prepared to intervene in significant matters, like acquisitions, and act as a safeguard when necessary. However, they tend to and should avoid getting entangled in operational details that could distract from their strategic focus.

The dynamic at the board level is delicate. Misplaced information can disrupt both the board's and the CEO's agendas. The perception that the CEO is limiting your access to the board is not necessarily about hiding information but about preventing unnecessary risks that could distract from the broader strategic issues the board needs to concentrate on.

Regarding the CEO's relationship with the board, while the board supervises the CEO, the CEO influences the board's operational level and tone. The CEO's ability to derive value and support from the board hinges on maintaining the board's focus at the necessary strategic level.

The board's responses and insights are shaped by the information and queries presented to them. High-level, strategic discussions prompt correspondingly strategic thinking, whereas detailed presentations may lead to detailed scrutiny. Thus, the CEO aims to use their time with the board strategically, ensuring discussions are aligned with the board's oversight and insight roles.

Seek to Understand Board Roles, Including the Role of Lead Director

The lead director plays a key role in setting the agenda and coordinating with other board members. They are often the CEO's first point of contact in a crisis or incident, jointly deciding with the CEO whether to call an emergency board meeting or take another course of action based on the situation.

The CEO is not obligated to follow every directive from the board. While it's crucial for the CEO to consider and respond to the board's advice, the board is not part of the company's management. Up to and including the risk of being fired, the CEO may act against the board's advice.

When preparing to address the board or discuss board engagement with your CEO or another leader, certain mindsets are critical. Adopting the wrong approach can signal that you're not ready for board engagement.

Here's a perspective shift to consider: Instead of seeking the board's assistance in decision-making or advocating for cybersecurity funding (which could lead to issues), position yourself as supporting the CEO, the board, and ultimately the company. This reframing emphasizes that your objective is to contribute positively to the board's ability to support the company, creating a stronger justification for your engagement with the board.

A common question I encounter is, "How do I secure a seat at the table? How can I gain visibility with the board or receive an invitation?" While securing a spot may seem crucial to you, the CEO views you as an extension of the company's management. The less you focus on personal visibility and the more you align with representing the company's management, the better your chances of engaging with the board.

When considering the agenda and topics of discussion, be prepared to adapt. You may enter the conversation with the CEO or their board-prep delegate with certain expectations, but be ready for these to shift significantly. For your contributions to be effective, they should align with and support the CEO's strategy and messages.

The board does not provide answers but rather perspectives. Approaching the board with the expectation of direct solutions can lead to disappointment. Instead, seeking their perspective or framing discussions to elicit their viewpoints without demanding specific decisions will likely yield more productive outcomes.

The most crucial point, which I've alluded to throughout, is that your readiness is not determined by your own assessment but by your leadership's perception.

Be thankful if you're asked to work through the CEO or a delegate when preparing to meet the board of

directors. Ask for this kind of advisor if you don't have one.

Right after your inside organizational coach, your board coach is the next most important member of your workplace circle of advisors.

Call me if you don't get a board coach, as it's a huge missed opportunity.

H: Implicit Bias Dimensions Among CISOs

Two dimensions tend to dominate how a CISO operates in their role.

Those two dimensions concern the following tendencies, mental models, or defaults:

- The spectrum from Protect Opportunity ("Opportunity Assumption") to Protect Assets ("Asset Assumption")
- The spectrum from Centralized Decisions (Often "CISO Knows Best") to Collaborative Decisions ("Collective Knows Best")

Put in a two-by-two box, the dimensions look like this:

	CISO DECISION	COLLECTIVE DECISION
PROTECTION ASSUMPTION	TYPE I • Arrogance: moderate • Alignment: low • Personality: enforcer • Range: short but fast	TYPE II • Arrogance: moderate • Alignment: moderate • Personality: expert • Range: long but tiresome
OPPORTUNITY ASSUMPTION	TYPE IV • Arrogance: high • Alignment: low-moderate • Personality: savior • Range: moderate but unreliable	TYPE III • Arrogance: negligible • Alignment: high • Personality: partner • Range: long with bonus of support

No cyber or cyber-risk models operate more in the opportunity assumption than the protection assumption, in much the same way scientific taxonomies don't solve for what kind of flowers are appropriate for a birthday or funeral.

Businesses generally primarily operate in the opportunity assumption space. If you continue to go to the business using asset protection models, you're not leaning into the business of the business, and the elusive seat at the table remains elusive.

You can't stay where you are and expect a new outcome.

I: Building Consensus

Progressively Prepare for the Conversations

The following techniques help you prepare for and sequence these important conversations:

1. Rubber Ducking: Talk your idea out loud to yourself, like programmers do when debugging code. This practice helps you identify and eliminate components driven by personal passion rather than relevance, refining your pitch for credibility.
2. Consult Confidants: Discuss your idea with trusted confidants who support your career and information security. They can help you identify potential challenges, refine your pitch, and navigate the decision-making process by providing insights into organizational mechanics and external factors.
3. Seek Likely Advocates: Move on to individuals who are likely to support your initiative. Use their feedback to further refine your approach and gather support.
4. Engage Challengers: When approaching those who might

challenge your idea, either ask directly for their support to foster an honest discussion or build a strong enough base of support so that their objections carry less weight. Both approaches have their risks and benefits.

During your information-gathering process, pay attention to what people like about your idea and address their concerns directly. Use positive feedback from previous conversations to strengthen your case with new stakeholders, demonstrating that your idea has broader support and addressing fears such as budget constraints or timeline issues.

> **Note:** Share accolades and endorsements as you engage with stakeholders. When moving from your likely advocates to those less supportive, mention positive feedback from respected colleagues: "I talked to Bob, who's known for being tough on these projects, and he liked this idea because . . ." This brings additional perspectives into the conversation, lending credibility and showing broader support.

Techniques to Build Consensus

Examples of how to build consensus effectively:

- Incorporating Feedback: "I spoke to Lisa. She had a couple of suggestions, which I incorporated, making her less likely to challenge and reducing potential extra work for others."
- Addressing Concerns: "Doug had an issue with section five that would negatively affect his group. We eliminated that issue without compromising our goals, so he's now on board."

- Highlighting Support: "Kai really liked a particular component, so we've made it a requirement. This ensures their support."

Benefits:

- Eases Worries: Stakeholders are less concerned about backlash as they see growing consensus.
- Reduces Resistance: With more people on board, individuals won't have to fight as hard for the idea.
- Looks Good: Success of the project reflects positively on those involved.
- Minimizes Stress: By addressing concerns, you prevent increasing their workload or anxiety.

Side Effects:

- New Directions: Through this process, you might decide the project doesn't make sense or needs a new direction, which boosts your credibility and political capital by showing you value others' input.

You Run Into Immovable Objects if You Aren't Looking for Them

An immovable object is an unrecognized barrier deeply ingrained in an organization's culture or operations. These also go by other terms, some of dubious or ambiguous use by modern standards, such as *sacred cow*, *golden child*, *kingmaker*, *rainmaker*, and *untouchable* or *chosen one*. Trying to overcome or move these can ruin careers. Yours.

Recognizing these unmovable objects is crucial to avoid pushing ideas or approaches that will be fiercely resisted or rejected.

This is a complex topic, but here's a quick overview:

- Recognize and Respect: Identify immovable objects to avoid futile battles and save your credibility and career.
- Conserve Energy: Adapting your plans based on feedback prevents wasted effort on unviable ideas, preserving your energy and passion for projects that can succeed.

Remember, involving others in your discussions builds credibility and touches many areas. People will let you know what to avoid while you're talking with them; these things are not advertised or posted on the intranet.

Work with the Underpinning Culture, Most of the Time

A wise adage tells us that "culture eats strategy for lunch." Leaders know this and are protective of the culture that enables them to succeed, and enables the company to operate in its own distinct ways. They know companies can win on culture. Acknowledging this helps you design acceptable approaches.

Each industry and each company has a culture and ethos.

Don't fight against the central tenets of how an organization operates to succeed, even if it makes it harder to protect the organization. Your notions of requiring "culture" change to achieve risk avoidance are generally of less importance than protecting the core ethos that enables operations and goal achievement.

If a marketing agency has a creative spirit and ethos that is central to how they work, you have to work with that as their CISO, and you're going to move more into monitoring and response. You're also likely to have to balance that culture within your own team. If a fintech start-up has a culture of breaking everything except the accounts and you came from banking, where the culture is "don't break anything ever," you're going to have to make a shift to the tech start-up culture, which is agile and risk taking.

The possibilities are endless. So you're likely to need help. See chapter 5 on building networks and support.

Index

Quick Reference Section

- Collaboration: 2.7, 5.1, Appendix G
- Leadership: 4.1, 5.8, 8.3
- Opportunity Framing: 1.6, 1.7, 5.1
- Resilience: 1.3, 8.3, 9.3
- Strategy: 1.5, 6.2-6.8, 7.3
- Trust: 2.1, 5.2, 5.8 (p. 115)

Business Mindsets

Fostering Innovation within Constraints

- Creating solutions under limited budgets or resources, 17, subsection 1.4
- Building a culture that views constraints as part of opportunities, 18, subsection 1.5

Credibility in Leadership

- Maintaining authority through strategic framing, 13, subsection 1.1
- Navigating perceptions of technical expertise, 13, subsection 1.2
- Building long-term trust with executive peers, 48, subsection 3.1

Linking Cybersecurity to Business Opportunities

- Framing cybersecurity as an enabler of growth, 15, subsection 1.3
- Aligning risk mitigation with business goals, 18, subsection 1.5
- Practical models for resource allocation, 207, Appendix C

Holistic Risk Framing

- Shifting from heat maps to opportunity-grounded narratives, 20, subsection 1.6
- Addressing risk tolerance within a business context, 22, subsection 1.7
- Challenges of isolating cybersecurity from broader risk, 23, subsection 1.7

Influence and Leadership

Navigating Power Dynamics

- Understanding implicit and explicit power structures, 31, subsection 2.2
- Persuading peers and stakeholders outside direct reporting lines, 37, subsection 2.5
- Leveraging credibility and trust to inspire collaboration, 38, subsection 2.5

Building Trust and Authority

- Using the trust equation: credibility, reliability, intimacy, 87, subsection 5.2
- Vulnerability in executive leadership, 70, subsection 4.5
- Setting aside your personal agenda, 95, subsection 5.5

Stakeholder Engagement Cornerstones

- Orienting to others' perspectives and professions, 67, subsection 4.3
- Understanding stakeholders' priorities, 94, subsection 5.4
- Structuring conversations to build shared goals, 94, subsection 5.5
- Understanding what leaders need from you, 104, subsection 5.8

Collaborative Problem-Solving

- Designing solutions with—not for—business leaders, subsection 1.4
- Using business input to strengthen a case, 35, subsection 2.5
- Using pre- and postmortems to foster alignment, 100, subsection 5.7

Strategy and Decision-Making

Principles for Effective Strategy

- Defining addressable challenges versus grievances, 120, subsection 6.1
- The integration of policy and action, 125, subsection 6.1
- Moving beyond plans, 136, subsection 6.3

Dynamic Adjustments in Leadership
- Adaptive models over rigid frameworks, 138, subsection 6.5
- Recognizing emergent strategies in a shifting landscape, 139, subsection 6.7
- Balancing short-term wins with long-term vision, 115, subsection 5.8

Business Impact Alignment
- Reframing cybersecurity risk as business decision support, 18, subsection 1.5
- Leveraging informal processes to drive impact, 22, subsection 1.7
- Co-defining the CISO role with stakeholders, 106, subsection 5.8

Scenario Planning and Simulation
- Using scenarios to anticipate challenges and opportunities, 99, subsection 5.7
- Discussing outcomes to inform better strategic decisions, 95, subsection 5.5

Strategic Delegation
- Empowering team members to make decisions within a strategic framework, 136, subsection 6.3
- Delegating without losing alignment with overall goals, 138, subsection 6.6

Personal and Professional Growth

Work-Life Balance
- Navigating boundaries between personal and professional roles, 108, subsection 5.8
- Strategies for maintaining energy and focus in demanding leadership positions, 160, subsection 7.2

Building a Legacy
- Defining long-term goals and the impact you want to leave behind, 163, subsection 7.3
- Guiding others to carry forward your vision, 166, subsection 7.4

Continuous Learning and Unlearning
- Staying adaptable through ongoing self-awareness, 176, subsection 8.3
- Identifying and letting go of outdated mindsets or practices, 187, subsection 9.2

Self-Coaching Methods
- Tools for turning experience into actionable insight, 170, subsection 8.1
- Using feedback loops to refine leadership behaviors, 176, subsection 8.3
- Overcoming resistance to personal growth, 185, subsection 9.2

Balancing Identity as a Business Executive
- Transitioning from technical expert to holistic leader, 48, subsection 3.1

- Letting go of "cyber-first" thinking to gain broader impact, 15, subsection 1.3
- Finding fulfillment in the evolving role of the CISO, 191, Conclusion

About the Author

Chris Brown is an executive coach to CISOs, a former CISO himself, and a recognized voice in cybersecurity leadership and strategy. He has spent decades shaping security strategy across Fortune 500 and mid-size businesses, guiding cybersecurity executives in building influence at the executive level while operating in harmony with business objectives.

As founder of *New Cyber Executive*, Chris works with cybersecurity leaders ready to move beyond heads-down execution and into the real work of executive leadership. Through his CISO Impact System, he coaches clients on communication, risk dynamics, and business-favored mindsets—helping them become trusted voices in the CEO's circle, in ways that transform how cybersecurity is experienced and viewed across the organization.

He's the author of *CISO Impact and Influence: Take the Lead and Nudge the World*, a guide for security executives ready to lead with more authority, clarity, and connection. His work has been featured in *Directors & Boards*, *Financial Times*, and *Fast Company*, and he leads workshops on executive readiness, strategy, and reputation.

Chris's work challenges old models of security leadership. He advocates for a relationship-driven, culturally aware, business-first approach—where CISOs don't just protect assets, but become essential business leaders.